Composition and Properties of Drilling and Completion Fluids

Composition and Properties of Drilling and Completion Fluids

Editors

Madhuri Parthi/Ambrish Kumar

Composition and Properties of Drilling and Completion Fluids

Edited by **Madhuri Parthi/Ambrish Kumar**

Printed in 2017

ISBN: 978-1-68117-345-0

Library of Congress Control Number: 2015939258

© 2016 by
SCITUS Academics LLC,
616, Corporate Way, Suite 2, 4766,
Valley Cottage, NY 10989

www.scitusacademics.com

Contents

Preface

Composition and Properties of Drilling and Completion Fluids has been updated and revised to incorporate new information on technology, economic, and political issues that have impacted the use of fluids to drill and complete oil and gas wells. With updated content on Completion Fluids and Reservoir Drilling Fluids, Health, Safety & Environment, Drilling Fluid Systems and Products, new fluid systems and additives from both chemical and engineering perspectives, Wellbore Stability, adding the new R&D on water-based muds, and with increased content on Equipment and Procedures for Evaluating Drilling Fluid Performance in light of the advent of digital technology and better manufacturing techniques, Composition and Properties of Drilling and Completion Fluids has been thoroughly updated to meet the drilling and completion engineer's needs.

Editor

Enhanced Viscosity of Aqueous Palygorskite Suspensions through Physical and Chemical Processing

Feng-shan Zhou[1], Tian-qi Li[1], Yun-hua Yan[2], Can Cao[1], Lin Zhou[1], and Yang Liu[1]

[1]School of Materials Science and Technology, China University of Geosciences (Beijing), Beijing 100083, China

[2]The Key Laboratory of Orogenic Belts and Crustal Evolution, School of Earth and Space Sciences, Peking University, Beijing 100871, China

ABSTRACT

Palygorskite has remarkable rheological properties and was used to increase the stability and viscosity of aqueous suspensions. The effects of different physical and chemical processing methods on the apparent viscosity and plastic viscosity of the palygorskite suspensions such as pressing, ultrasound scattering, acidification, and chemical additives have been released. The pressing and ultrasound scattering indicated that the dispersed state of palygorskite could be increased effectively after treatment, and the apparent viscosity of treated-palygorskite samples increased almost 2-3 times compared to that of before. The viscosity of the acid-treated palygorskite suspension was not increased. The viscosity increased with the content of bentonite in the mixture of bentonite and palygorskite in fresh water. It seemed to be not worthy to add a certain amount of bentonite to palygorskite in order to enhance viscosity and vice versa. Chemical additives appeared to have good effects on the rheological behavior of palygorskite suspension. Magnesium oxide revealed great contribution to viscosity enhancement. The main mechanism was the electrostatic attractive interaction between magnesium oxide particles with positive charges and the palygorskite rods with negative charges. This interacted force has an impact on the structural inversion of palygorskite rods and even caused the reinforcing of flocculation.

INTRODUCTION

Clay minerals have remarkable rheological properties and are used to increase the stability and viscosity of flowing suspensions. They tend to form gel-like structures at low solid contents [1]. This property is of great importance in a very wide range of applications for drilling fluids, paints, liquid fertilizers, wild-fire suppressants, foundry coatings, animal flowing feeds, molecular sieve binders, and a lot of aqueous suspensions in which rheological properties play a significant role [1–5].

Palygorskite forms gel structures in fresh and salt water by establishing a lattice structure of particles connected through hydrogen bonds. In the drilling industry, these properties enable the clay suspension to suspend the large dense particles of the drilling cuttings and require relatively low pump power during water circulation [1, 7]. Palygorskite, unlike bentonite, will form gel structures in salt water and is used in special salt-water drilling mud for drilling formations contaminated with salt [8]. Palygorskite particles can be considered as charged particles with zones of positive (+) and negative (−) charges. It is the bonding of these alternating charges that allows them to form gel suspensions in salt and fresh water.

Although most clay minerals form stable and viscous suspensions when dispersed in water, the mechanisms of gel formation for each clay mineral differ because of their unique structures, particle size and shape, and composition [9, 10].

Unlike the swelling clay minerals such as montmorillonite, palygorskite as a fibrous nonswelling clay mineral, the fibre length and number of silanol groups on the surface of the fibre play an important role in aggregating fibres together [11] and forming a random network that entraps water and increases viscosity [12].

Neaman and Singer [13–15] systematically studied the rheological properties of six palygorskite samples, used them as a kind of common thixotropic modifier in aqueous suspensions, and focused on the influence factors including the ratio of crystal length to diameter, concentration of sodium chloride (NaCl), and pH. The effects of pressing modification and adding magnesium oxide (MgO) [16, 17], ultrasound scattering [18, 19], acidification [6, 20], and even the mixed palygorskite-bentonite suspensions [21] on the rheological properties of palygorskite suspensions also have been studied.

High-pressure homogenization process with solvent and electrolytes with dispersion properties of palygorskite were investigated in detail by Xu et al. [22–24]. They dispersed the natural palygorskite in six solvents including distilled water, methanol, ethanol, isopropanol, dimethyl formamide, and dimethyl sulfoxide and then carried out high-pressure homogenization. They confirmed that colloidal stability and suspension viscosity were affected by the solvent nature, and a much higher viscosity was obtained by dispersing palygorskite in isopropanol, but the good colloidal stability was obtained in dimethyl sulfoxide (DMSO) solvent. A series of palygorskite samples modified with inorganic potassium electrolytes including KCl, KBr, KI, KH_2PO_4, $KHSO_4$, K_2HPO_4, K_2SO_4, and K_3PO_4 were prepared with the aid of high-pressure homogenization. A stable suspension was obtained when palygorskite was dispersed in K_2SO_4 solutions. Because the requirement of the viscosity is more than the stability for palygorskite suspension, obviously, high prices and toxic solvents in their works would be limiting the value of industrial applications.

However, no one had systematically studied the influence of viscosity and the methods of enhanced viscosity for palygorskite gel. The purpose of the present work is to study the effects of different physical and chemical processing methods, such as pressing, ultrasound scattering, acidification, and chemical additives, on the apparent viscosity and plastic viscosity of the aqueous palygorskite suspensions.

MATERIALS AND METHODS

Materials

Palygorskite mineral sample with purity greater than 95% was received from Mingguang Palygorskite Mining Co., Ltd. (Anhui, China). The average length of the palygorskite rods is around 1 μm, and the average aspect ratio is about 20. The chemical compositions, the physical properties, and the theoretical crystal structural formula of the palygorskite sample in present work are listed in Table 1.

Table 1: The chemical compositions, the physical properties, and the theoretical crystal structural formula of the palygorskite sample in present work

Components	SiO_2	Al_2O_3	TiO_2	Fe_2O_3	FeO	MgO	CaO	Na_2O	K_2O	P_2O_5	SO_3
Content/ wt%	64.89	12.95	1.26	8.19	0.16	9.11	1.48	0.07	1.33	0.43	0.02
Crystal structural formula: $(Mg_{0.81}Al_{0.73}Fe^{3+}_{0.36}Ca_{0.09}Ti_{0.06})(Si_{3.83}Al_{0.17})O_{10}(OH)\cdot 4H_2O$											
Cation exchange capacity (CEC): 49.8 meq/100 g											
Specific surface area: 462.0 m²/g											

There are two bentonite mineral samples used in this work. One bentonite sample was obtained from Sinopec Shengli Oilfield Co., Ltd. (Shandong, China). Another bentonite containing palygorskite sample from Iraq Anbar was obtained from Beijing Taihua Bentonite Science & Technology Development Co., Ltd. (Beijing, China). Industrial grade magnesium oxide (also named calcined magnesia; light-burned magnesia) (MgO) and magnesium hydroxide ($Mg(OH)_2$) samples were obtained from Dandong Yilong High Science & Technological Materials Co., Ltd. (Liaoning, China).

Instruments

The experiment of pressing palygorskite was conducted on Jinniu JL-80 vertical grinder (1.5 KW; Hualian Industry Co., Ltd.) (Beijing, China). The ultrasonic dispersion test was carried out by JY92-IIDN ultrasonic cell crusher (20–24 KHz, 650 W; Ningbo Scientz Biotechnology Co., Ltd.) (Zhejiang, China). The modified palygorskite was filtrated by SHB-III water circulation pumps (180 W; Xi'an Taikang Biotechnology Co., Ltd.) (Shaanxi, China) and dried by a 101-3 electric blast drying oven (300°C maximum; Shanghai Rolling-gen Equipment Co., Ltd.) (Shanghai, China). The gelation of samples dispersion was prepared after being stirred by GJ-2S digital display high-speed agitator (180 W, 4000–11000 rpm). The rheological parameters of suspensions were conducted by ZNN-D6A six-speed rotary viscometer (speed: 3, 6, 100, 200, 300, and 600 rpm, viscosity range: 0–300 mPa·s). Both of the latter instruments were manufactured by Qingdao Haitongda Special Instrument Co., Ltd. (Shandong, China).

Methods

Palygorskite Modification

Pressing: A certain amount of palygorskite and tap-water was mixed and pressed by a vertical extruder. After that, the palygorskite was collected, dried, and ground.

Ultrasound Scattering: 25 g palygorskite and a certain amount of tap-water were added into an 80 mL beaker, stirred, and then loaded onto the platform of an ultrasonic cell crusher. The time of ultrasound scattering on the platform of ultrasonic cell crusher was 10 min and repeated 2-3 times to make a better dispersion.

Acidification: The clay mineral was treated with hydrochloric acid at a concentration of 2 mol/L by liquid and a solid ratio of 10 to 1 in the flask, under mechanical stirring (550 rpm) in dispersion at room temperature for 1 h. Then the sample was filtrated, followed by washing with distilled water until a pH value 3-4 was reached.

Chemical Additives: Certain amounts of additives were added into 6.4 w/v% palygorskite dispersions. The dosage of chemical additives (%) in the tables and the figures was the ratio of mass between chemicals and palygorskite.

The dispersions were stirred mechanically at 8000 rpm for 20 min at room temperature and hydration was conducted for 24 h. The term of hydration was an ageing process of water penetrating the interlayer spaces and concomitant adsorption with the clay swelling and colloidization.

Rheological Parameters Measurement

Darley and George [8] concluded the common composition and properties of drilling and completion fluids. According to the American Petroleum Institute (API) recommended practice (2009) [25], the parameters of the palygorskite gel in the drilling fluid suspension samples were prepared and measured under the specification and standard procedures. The viscosity and gel strength of the modified gels were tested by a rotating viscometer (ZNN-D$_6$S). The hydrated gels underwent mechanical stirring at 8000 rpm for a further 20 min. This preparation step before measuring the viscosity was to make the dispersion even and flowing. And then the viscosity was measured at different shear rates (different stirring velocity).

Rheological Theory

According to Bingham-plastic model, the rheological parameters, including AV (apparent viscosity), PV (plastic viscosity), YP (yield point), and RYP (ratio of yield and plastic viscosity), were calculated with the dial readings of 300 rpm and 600 rpm using the following formulas according to the API recommended practice of standard procedures [25]:

$$AV = 0.5\theta_{600} \; (mPa \cdot s)$$

$$PV = \theta_{600} - \theta_{300} \; (mPa \cdot s)$$

$$YP = 0.511 \left(\theta_{300} - PV \right) \; (Pa)$$

$$RYP = \frac{YP}{PV} \; (Pa/mPa \cdot s),$$

(1)

where θ_{600} (dia) was the dial reading of rotating viscometer at 600 rpm and q_{300} (dia) was the dial reading of rotating viscometer at 300 rpm.

Microscopic Examination

The morphology of the palygorskite and modified palygorskite with additives was observed in a Quanta 200FEG environmental scanning electron microscope (SEM). All of the raw minerals were firstly made to powder samples, which were dried from dilute 0.2% dispersions before the SEM examination. The modified palygorskite gel samples were also dried out from the same concentration before SEM measurement.

RESULTS AND DISCUSSION

Pressing Effect

Pressing as an effective way to break up close and compact bulks of natural palygorskite rods could more easily disperse palygorskite fibers in water. Experimental measurement of viscosity of pressed palygorskite showed that the apparent viscosity increased from 19.5 mPa·s to 42.5 mPa·s (Table 2).

Table 2: The effect of pressing on the rheological parameters of palygorskite suspension

Samples of palygorskite	Rheological parameters					
	θ_{600}(dia)	q_{300}(dia)	AV (mPa·s)	PV (mPa·s)	YP (Pa)	YP/PV (Pa/ mPa·s)
Unpressing	39	35	19.5	4.0	15.84	3.96
Unpressing with 1% MgO	59	44	29.5	15.0	14.82	2.00
Pressing without MgO	48	41	24.0	7.0	17.37	2.48
Pressing with 1% MgO	85	75	42.5	10.0	33.22	3.32

Pressing can be applied as a useful approach to enhance better dispersion of palygorskite particles, especially for raw palygorskite aggregates. However, in our opinion, compared to effective viscosity enlarging by using small amount of chemical, the pressing with the inefficient and high cost was not a suitable technique for enhanced viscosity of palygorskite suspension in some applications with low added value such as drilling fluids.

Ultrasound Scattering Effect

Song et al. [18] found that after treatment with ultrasound the palygorskite crystal bundles were crushed into crystal needles, generating palygorskite nanoparticles. What is more, Zhao et al. [19] studied the dispersion of palygorskite in a polypropylene matrix, and their SEM and TEM analysis results showed that ultrasonic oscillation promoted the dispersion of palygorskite particles (Figure 1).

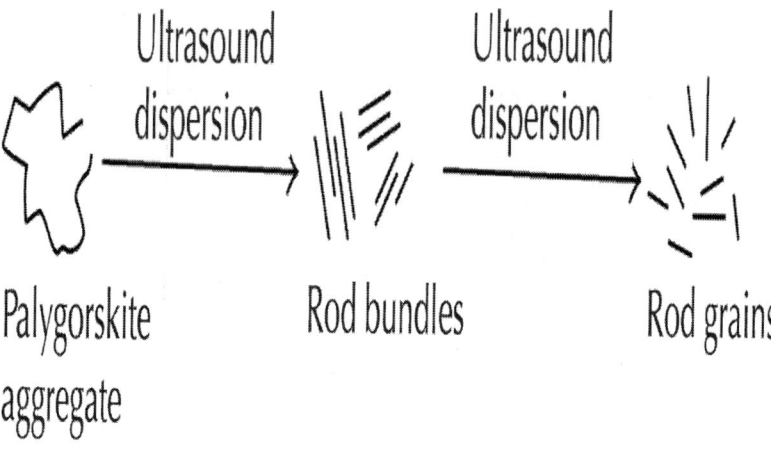

Figure 1: The mechanism of ultrasound scattering dispersion for palygorskite aggregates.

Under the condition of ultrasound dispersion, the apparent viscosity was raised rapidly from 19.5 mPa·s to 54 mPa·s (Table 3). The enhancement of viscosity was related to the fact that ultrasonic cavitation could cause local high temperature and high pressure. The shock wave and microjet in dispersion brought about intense collisions of palygorskite aggregates in aqueous suspensions just like the same dispersion mechanism in polypropylene matrix described in Figure 1.

Table 3: The effect of ultrasound scattering on the rheological parameters of palygorskite suspension

Samples of palygorskite	Rheological parameters					
	θ_{600}(dia)	q_{300}(dia)	AV (mPa·s)	PV (mPa·s)	YP (Pa)	YP/PV (Pa/ mPa·s)
No ultrasound scattering	39	35	19.5	4	15.84	3.96
Ultrasound scattering	108	92	54.0	16	38.84	2.43

Acidification Effect

Many practices using acid to purify palygorskite have been reported in the literature. For example, Neaman and Singer [20] used acid to remove carbonates and other cement impurities. Other researchers used acidification to break up the cluster of closely bound fibers to increase specific surface area for good dispersion and absorption. The raw material in the experiment was of high purity. Octahedral cations dissolved and crystal structure was even changed because of the high concentration of acid and reaction time. Chen et al. [6] investigated the structural changes of palygorskite with reaction to acid; their results indicated that dissolution of octahedral reactions increased with an increase in acid concentration and reaction time. When octahedral cations were dissolved completely, the final product was mesoporous amorphous silica-fiber (Figure 2).

(a)

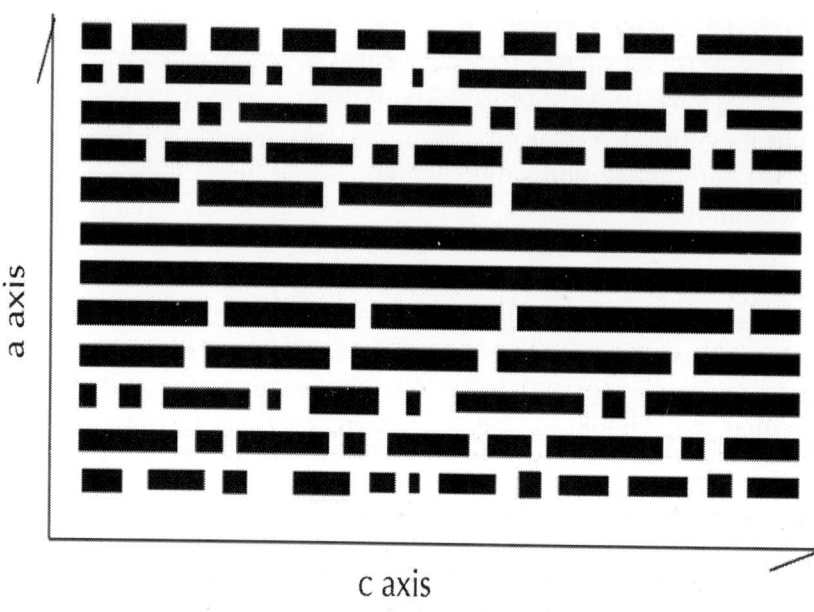

(b)

Figure 2: The channel structure change of palygorskite with acidification treatment [6]. (a) The original channel structure of palygorskite. (b) The channel structure of palygorskite after acidification with hydrochloric acid.

The study showed that the gel of the acidified palygorskite in dispersion came to serious sedimentation after hydration. Initially, with the ratio of acid to palygorskite being 10 to 1 and concentration of HCl being 2 mol/L, the apparent viscosity value was only 5 mPa·s. By reducing the concentration and amount of acid solution (weight ratio of acid to palygorskite was 4 to 1; concentration of HCl was 1 mol/L), the AV still measured only 10 mPa·s. Obviously, acidizing palygorskite viscosity was not enhanced in the present work; the opposite is the case.

Effects of Bentonite

Several studies have been carried out in the past to understand the rheological properties of standard clays [9, 10, 13–15, 26–28].

However, there are a lot of works on the rheological properties of mixed clay suspensions in recent years. The influence of montmorillonite addition on the rheological behaviour of palygorskite suspensions was investigated by Neaman and Singer [13–15]. Rheological properties of palygorskite-bentonite mixed clay suspensions were studied by Chemeda et al. [21].

The limited information is considered important because most clay used such as in drilling fluid applications usually contains more than one type of clay minerals along with nonclay minerals. For example, palygorskite occurs in association with smectite in most of the known world palygorskite deposits [6]. Therefore it is worthwhile to understand the rheological behavior of suspensions containing mixtures of clay minerals.

The most difference of rheological properties between palygorskite and bentonite was that palygorskite can be used in fresh water and salt water, but the bentonite is only used in fresh water. Figure 3 shows the effects of bentonite addition on the viscosity of bentonite-palygorskite mixture. The viscosity of the mixture increased with the increase of content of bentonite in mixture in fresh water. But the viscosity of the mixture decreased with the increase of content of bentonite in mixture in salt water because of the poor salt tolerance of bentonite. Taking into account the effect, nature, and the price ratio, it was not worthy to add a certain amount of bentonite to palygorskite, because the palygorskite is normally used in salt water condition. For the same reason, it was also worthless to add a certain amount of palygorskite to bentonite.

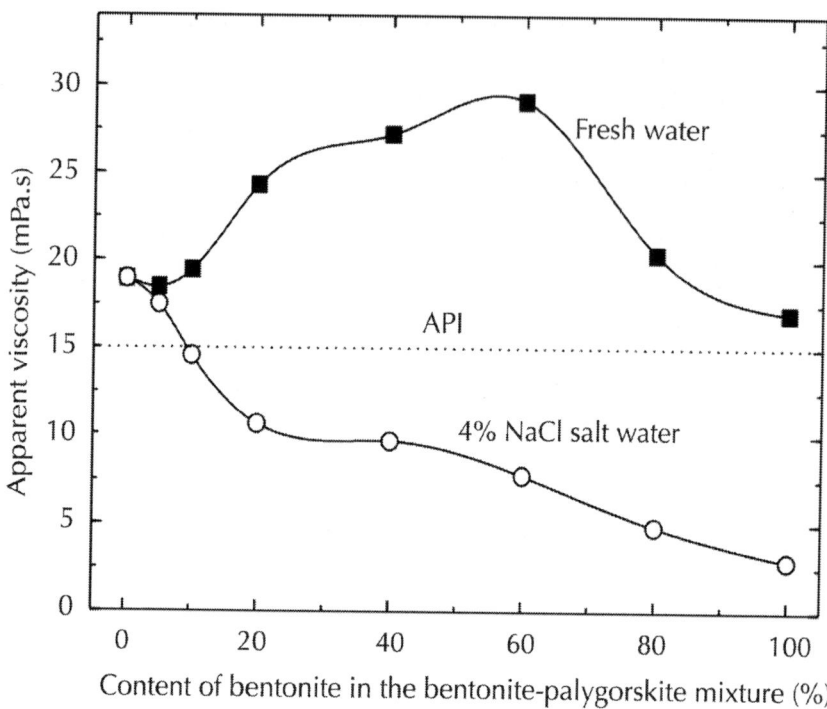

Figure 3: The effects of bentonite addition on the viscosity of suspension of bentonite-palygorskite mixture in fresh water and salt water.

In contrast to the admixture of bentonite and palygorskite, some kinds of natural coexisting bentonite-palygorskite clay mixture would have a very high viscosity (Figure 4), because their random network structures were formed more easily which entrapped water and increased viscosity. But for some others, the same viscosity behavior does not appear [6]; the real reason is still unknown.

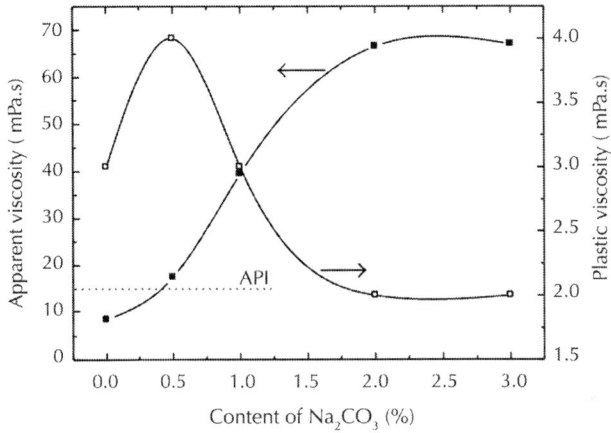

(a)

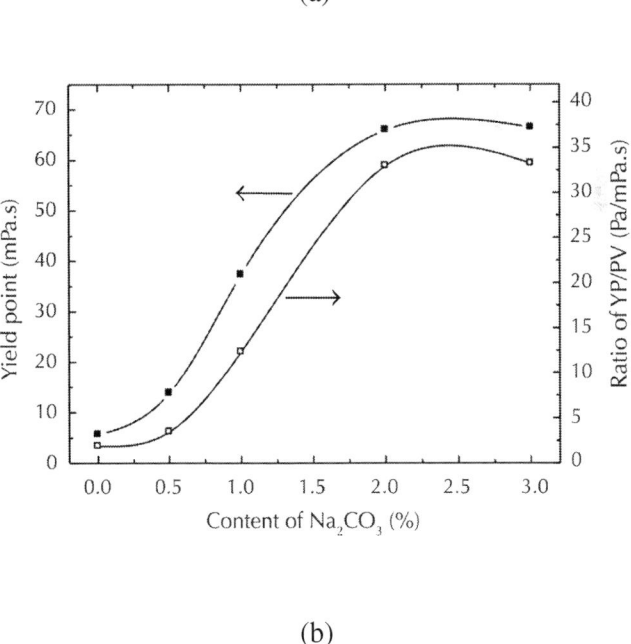

(b)

Figure 4: The viscosity of a natural coexisting clay mixture sample with 7% palygorskite and 57% sodium-calcium based hybrid bentonite from Iraq Anbar. (a) The apparent viscosity and plastic viscosity of the natural coexisting clay mixture. (b) The yield point and ratio of YP/PV of the natural coexisting clay mixture.

Effects of Magnesium Oxide

The experimental measurements of samples with chemical additives (Figure 5) exhibited that the sample added MgO showed an increased viscosity value with lower MgO content. This higher viscosity value exhibited better cuttings suspension and carrying capacity in drilling fluids.

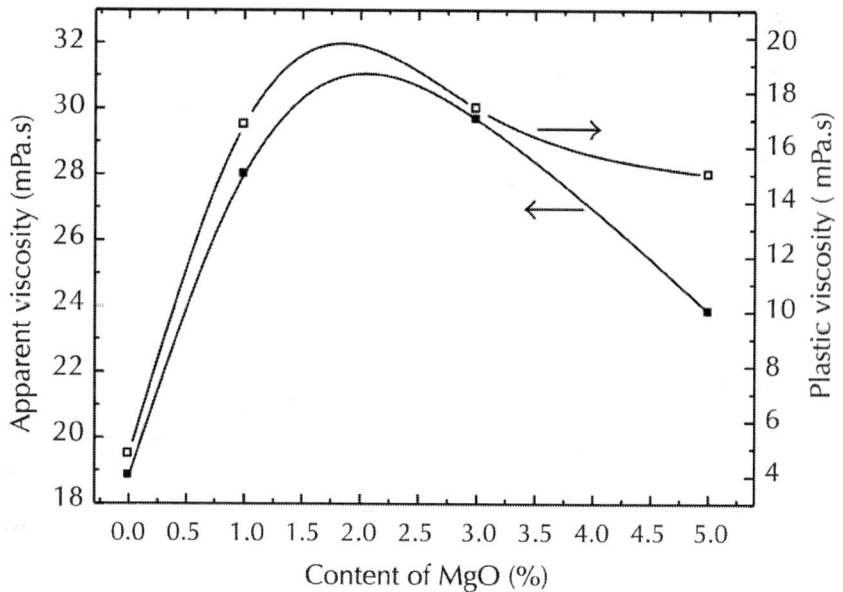

(a)

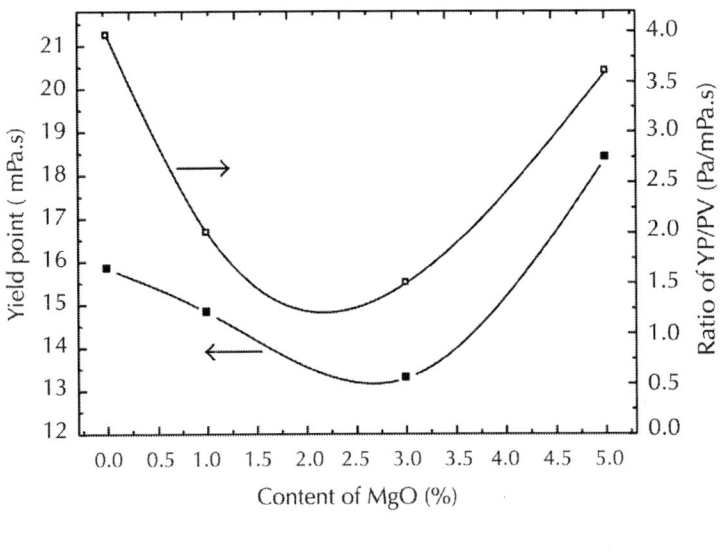

(b)

Figure 5: The effects of MgO on the rheological parameters of palygorskite suspension. (a) The apparent viscosity and plastic viscosity of palygorskite suspension. (b) The yield point and ratio of YP/PV of palygorskite suspension.

When MgO particles were added to water, the reaction happened as follows:

$$MgO + H_2O \rightleftharpoons Mg(OH)_2 \rightleftharpoons Mg^{2+} + 2OH^-$$

(2)

The cation exchange ability of Mg^{2+} was better than Na^+. The Mg^{2+} entered into the channels of clay mineral particles and caused shrinkage of the electrical double layer. The shrinkage of the electrical double layer easily formed face-face aggregation. At the same time, the absorbed Mg^{2+} bridged edge and face formed edge-edge and edge-face flocculation.

As already stated above, the PV reflected the internal friction of suspended particles, the liquid phase, and their interface. Flocculation reinforced the suspension network structure with expression of an increase on viscosity.

SEM micrographs of the palygorskite with Mg(OH)$_2$ and MgO (Figure 6) revealed that the palygorskite had a fibrous morphology and that Mg(OH)$_2$ and MgO particles with positive charge dispersed in the palygorskite scaffolding structure with negative charge. The electrostatic attractive interaction also reinforced the palygorskite structure, confirming the increase in viscosity.

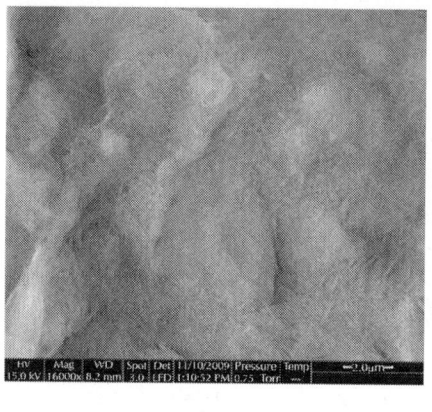

(a)

(b)

Figure 6: The change of scanning electron microscope photograph of palygorskite processed with magnesium oxide. (a) SEM micrograph of original unmodified palygorskite. (b) SEM micrograph of palygorskite modified adding MgO.

CONCLUSIONS

The results of the pressing and ultrasound scattering effect studies indicated that the dispersed state and increasing viscosity of clay mineral gel could be adjusted by the two methods effectively. The mechanisms were that pressing broke up close and compact palygorskite rods clusters, and ultrasonic cavitations caused intense collisions within the palygorskite aggregate. Consequently, pressing and ultrasound scattering could be used as useful modification methods for improving the viscosity of the aqueous suspension of palygorskite.

The acidification effect would not increase the viscosity of palygorskite, perhaps because the high concentration acidification was harmful for gelation and dispersion of palygorskite.

The viscosity of the mixture of bentonite and palygorskite increased with the increase of content of bentonite in fresh water. But the viscosity of the mixture decreased with the increase of content of bentonite in salt water because of the poor salt tolerance of bentonite. It seemed to be not worthy to add a certain amount of bentonite to palygorskite in order to enhance viscosity.

Chemical additives showed good effects on the rheological and thixotropic behavior of palygorskite suspension. The results of adding MgO revealed that the contribution of MgO to viscosity caused the reinforcing of flocculation. Furthermore, the analysis showed that electrostatic attractive interaction between MgO particles dispersed in the scaffolding structure with positive charges and the palygorskite rods with negative charges had impact on the inversion of palygorskite rods configuration. In drilling applications, this higher viscosity value will provide better cuttings suspension and carrying capacity.

ACKNOWLEDGMENT

Thanks are due to Dr. Susan Turner (Brisbane) for the helpful comments on the paper and for improving the English language.

REFERENCES

1.   P. F. Luckham and S. Rossi, "The colloidal and rheological properties of bentonite suspensions,"Advances in Colloid and Interface Science, vol. 82, no. 1–3, pp. 43–92, 1999.

2.   N. P. Chafe and J. R. de Bruyn, "Drag and relaxation in a bentonite clay suspension," Journal of Non-Newtonian Fluid Mechanics, vol. 131, no. 1–3, pp. 44–52, 2005.

3.   E. Galan, "Properties and applications of palygorskite-sepiolite clays," Clay Minerals, vol. 31, no. 4, pp. 443–453, 1996. ·

4.   E. Galan and A. Singer, Developments in Palygorskite-Sepiolite Research, A New Outlook on these Nanomaterials, vol. 3 of Developments in Clay Science, Elsevier, New York, NY, USA, 2011.

5.   H. H. Murray, "Occurrences, processing and application of kaolins, bentonites, palygorskite-sepiolite, and common clays," in Developments in Clay Science, vol. 2 of Applied Clay Mineralogy, Elsevier, 2007.

6.   T. H. Chen, Y. L. Feng, and X. Shi, "Study on products and structural changes of reaction of palygorskite with acid," Journal of the Chinese Ceramic Society, vol. 31, no. 10, pp. 959–964, 2004.

7.   V. C. Kelessidis, C. Tsamantaki, and P. Dalamarinis, "Effect of pH and electrolyte on the rheology of aqueous Wyoming bentonite dispersions," Applied Clay Science, vol. 38, no. 1-2, pp. 86–96, 2007.

8.   H. C. H. Darley and R. G. George, Composition and Properties of Drilling and Completion Fluids, Gulf Professional, Houston, Tex, USA, 6th edition, 2011.

9.   H. Heller and R. Keren, "Rheology of Na-rich montmorillonite suspension as affected by electrolyte concentration and shear rate," Clays and Clay Minerals, vol. 49, no. 4, pp. 286–291, 2001. ·

10.   E. Paineau, L. J. Michot, I. Bihannic, and C. Baravian, "Aqueous suspensions of natural swelling clay minerals. 2. Rheological characterization," Langmuir, vol. 27, no. 12, pp. 7806–7819, 2011.

11. T. C. Simonton, S. Komarneni, and R. Roy, "Gelling properties of sepiolite versus montmorillonite,"Applied Clay Science, vol. 3, no. 2, pp. 165–176, 1988. · ·

12. G. E. Christidis, P. Katsiki, A. Pratikakis, and G. Kacandes, "Rheological properties of palygorskite—smectite suspensions from the Ventzia basin, W. Macedonia, Greece," Bulletin of the Geological Society of Greece, vol. 43, pp. 2562–2569, 2011.

13. A. Neaman and A. Singer, "Kinetics of hydrolysis of some palygotskite-containing soil clays in dilute salt solutions," Clays and Clay Minerals, vol. 48, no. 6, pp. 708–712, 2000.

14. A. Neaman and A. Singer, "Rheology of mixed palygorskite—montmorillonite suspensions," Clays and Clay Minerals, vol. 48, no. 6, pp. 713–715, 2000. ·

15. A. Neaman and A. Singer, "Rheological properties of aqueous suspensions of palygorskite," Soil Science Society of America Journal, vol. 64, no. 1, pp. 427–436, 2000. ·

16. J. Zhou, L. J. Liu, N. Liu, and X. F. Liu, "Effects of $Mg(OH)_2$ and MgO on rheological behavior of attapulgite clay-water suspensions," Journal of Hefei University of Technology, no. 6, pp. 58–63, 1999.

17. J. Zhou, N. Liu, Y. Li, and Y. J. Ma, "Microscopic structure characteristics of attapulgite," Bulletin of the Chinese Ceramic Society, vol. 18, no. 6, pp. 50–55, 1999.

18. R. F. Song, L. Y. Yang, J. Sheng, N. X. Shen, and T. W. Kang, "The surface modification and characterization of nano-attapulgite," Bulletin of the Chinese Ceramic Society, vol. 22, no. 3, pp. 36–39, 2003.

19. L. Zhao, Q. Du, G. Jiang, and S. Guo, "Attapulgite and ultrasonic oscillation induced crystallization behavior of polypropylene," Journal of Polymer Science B: Polymer Physics, vol. 45, no. 16, pp. 2300–2308, 2007.

20. A. Neaman and A. Singer, "Possible use of the Sacalum (Yucatan) palygorskite as drilling muds,"Applied Clay Science, vol. 25, no. 1-2, pp. 121–124, 2004. · ·

21. Y. C. Chemeda, G. E. Christidis, N. M. Tauhid-Khan, E. Koutsopoulou, V. Hatzistamou, and V. Kelessidis, "Rheological properties of palygorskite—bentonite and sepiolite—bentonite

mixed clay suspensions," Applied Clay Science, vol. 90, pp. 165–174, 2000.

22. J. Xu, W. Wang, and A. Wang, "Effects of solvent treatment and high-pressure homogenization process on dispersion properties of palygorskite," Powder Technology, vol. 235, pp. 652–660, 2013.

23. J. Xu, W. Wang, and A. Wang, "Superior dispersion properties of palygorskite in dimethyl sulfoxide via high-pressure homogenization process," Applied Clay Science, vol. 86, pp. 174–178, 2013.

24. J. Xu, W. Wang, and A. Wang, "Influence of anions on the electrokinetic and colloidal properties of palygorskite clay via high-pressure homogenization," Journal of Chemical and Engineering Data, vol. 58, no. 3, pp. 764–772, 2013.

25. ANSI/API Recommended Practice 13B-1, Recommended Practice for Field Testing Water-Based Drilling Fluids, American Petroleum Institute, 4th edition, 2009.

26. S. Abend and G. Lagaly, "Sol-gel transitions of sodium montmorillonite dispersions," Applied Clay Science, vol. 16, no. 3-4, pp. 201–227, 2000.

27. L. V. Amorim, C. M. Gomes, H. L. Lira, K. B. Franca, and H. C. Ferreira, "Bentonites from Boa Vista, Brazil: physical, mineralogical and rheological properties," Materials Research, vol. 7, no. 4, pp. 583–593, 2004.

28. G. E. Christidis, A. E. Blum, and D. D. Eberl, "Influence of layer charge and charge distribution of smectites on the flow behaviour and swelling of bentonites," Applied Clay Science, vol. 34, no. 1-4, pp. 125–138, 2006.

Chapter 2

Effect of Nitric Acid on the Low Fluorescing Performance of Drilling Fluid Lubricant Based Animal and Vegetable Oils

Feng-shan Zhou[1], Zheng-qiang Xiong[1,2], Bao-lin Cui[1], Feng-bao Liu[1,3], Guang-huan Li[4], Jin-ran Wei[4], and Hua Cui[4]

[1]School of Materials Science and Technology, China University of Geosciences, Beijing 100083, China

[2]Beijing Institute of Exploration Engineering, Beijing 100083, China

[3]Tabei Exploratory & Development Department, PetroChina Tarim Oilfield Company, Korla 841000, China

[4]Drilling Fluid Technology Service Company, CNPC Bohai Drilling Engineering Ltd., Tianjin 300280, China

ABSTRACT

After synthesis of mixed fatty acid triethanolamine ester surfactant based on animal and vegetable mixed oils, the reaction solution was added into 4% (wt/wt) liquid nitric acid or 9% (wt/wt) solid nitric acid as eliminating fluorescent agent continuing to react from 1 to 2 hours. The low fluorescence lubricant named E167 for drilling fluid was prepared, in which maximum fluorescence intensity (F_{max}) was less than 10 in three-dimensional fluorescence spectra of excitation wavelength range. When the E167 was added into fresh water based drilling fluid at the dosage of 0.5% (wt/wt), the sticking coefficient reduced rate (ΔK_f) is 78% and the extreme pressure (E-P) friction coefficient reduced rate (Δf) is 79%. In the case of 4% brine mud with 0.5% (wt/wt) E167 in it, the ΔK_f and Δf are 75% and 62%, respectively. After the hot rolling ageing test 180° C × 16 h with the E167 was added into fresh water based drilling fluid at the dosage of 1% (wt/wt), the ΔK_f and Δf are greater than 70%, which shows a much better lubrication properties of strong resistance to high temperature. The fresh water based drilling fluid which contains 1% (wt/wt) E167 is almost nonfoaming even after hot rolling ageing 120° C × 16 h.

INTRODUCTION

In the process of oil and gas drilling, in order to reduce the friction between drilling string and borehole, drilling string and casing, together with reducing the drill string torque and tripping resistance, lubricant often need to be added, thus avoiding sticking accident and improving the drilling speed.

The most commonly used liquid lubricants for drilling fluid are mineral oils and vegetable oils. The mineral oils are difficult to biodegrade, which causes serious pollution problems to the environment, and the high grade of fluorescence is unfavorable for geological logging. In the case of vegetable oils, with properties of low toxicity, good biodegradability, resource renewability, and low fluorescence level, it is a kind of lubricant for environment friendly using as drilling fluid with promising application [1–7].

There are at least three problems in unmodified vegetable oils [8–12]: (1) the vegetable oils have low thermal stability in the process of hydrolysis, which are easy to saponify in alkaline environment, producing bubble of anion surfactant; (2) the adsorption consumption of water-soluble anion surfactant in the debris and borehole formation is larger than that of oil-soluble fat, which means that the consumption of lubricant after saponification is faster; namely, the lubricant after saponification is not durable; (3) with poor oxidation stability, the oil is easy to deteriorate resulting in stinking of drilling fluid, reducing and even losing the lubricating property. Therefore, chemical modification of vegetable oil is needed. The methods involved include hydrogenation [13], esterification [14, 15], and epoxidation [16].

The esterification modification of oil is studied in this paper using mixed oil as raw material, and the unsaturated bond of oil molecule was broken in the presence of nitric acid, and then a kind of low fluorescence lubricant with excellent lubricity was obtained.

PREPARATION AND PERFORMANCE EVALUATION OF LOW FLUORESCENCE LUBRICANT

Materials and Apparatus

The animal oil and vegetable oil were purchased, and solid nitric acid was prepared in our group. The purity of acid catalyst, triethanolamine, and nitric acid is all CP and GC for n-hexane.

The apparatus for measuring the adhesion coefficient was received Qingdao Haitongda Special Instruments Company; extreme pressure (E-P) lubrication device, OFI, USA; F-4600 fluorescence spectrophotometer, Hitachi, Japan; ZNN-D6S-six speed rotational viscometer and ZNS-2A-Low Pressure Filter Press, Qing Dao Haitongda Special Instruments Company; Infrared Spectrometer, Spectrum 100, Perkin Elmer, USA.

Preparation of Low Fluorescence Lubricant E167

Certain amounts of animal oil, vegetable oil, and acid catalyst were added into a three-necked flask equipped with a magnetic stirring bar, thermometer, and reflux condenser. The mixture was heated to corresponding temperature under stirring in given time, obtaining the ordinary lubricants without eliminating fluorescence. Then, the right amount of agent for eliminating fluorescence was added and the reaction continues for a certain time. Then, the target product was obtained.

Evaluation of Fluorescence Properties

At present, the UV visual for grading analysis method is commonly used for evaluating the fluorescent of drilling fluid lubricant. However, the method has the following problems [17]. (1) The wavelength emitted from UV lamp is around 365 nm, which is not enough to excite the fluorescence with wavelength lower than 365 nm. (2) Technically, it is difficult to accurately distinguish the fluorescence below 7 grades, using fluorescence logging instrument for naked-eye observation. (3) This method is only qualitative, but not quantitative, and with poor reproducibility. Therefore, according to the method described in patent by Patel [18], the fluorescence intensity of the lubricant F_{max} was quantitatively measured by the three-dimensional fluorescence spectroscopy, and the three-dimensional fluorescence spectra were obtained according to the optimal excitation wavelength and emission wavelength of EM. Methods and parameters are as follows: mix the test sample with n-hexane according to the mass ratio 1 : 400 and then measure the fluorescence properties with fluorescence spectrophotometer; both excitation and emission slit widths were fixed at 2.5 nm, and scan rate was selected at 1200 nm min^{-1} and PMT voltage at 700 V.

Evaluation of Lubricity

The adhesion coefficient reduced rate (ΔK_f) and lubrication coefficient reduced rate (Δf) are evaluation indicators of lubricity. The evaluation methods are according to drilling fluid liquid lubricant technical indicators Q/SY 1088-2007 [19].

The Evaluation of Rheological Properties

The rheological parameters apparent viscosity (AV), plastic viscosity (PV), yield point (YP), and filtration at the room temperature (FL_{API}) were

investigated according to the first part of the field testing of drilling
GB-T 16783.1-2006 [20].

Evaluation of Foamability

The method of foamability evaluation was performed referring to
the technical requirements and analytical methods of drilling fluid
lubricant Q/SY TZ 0022-2000 [21].

RESULTS AND DISCUSSION

The Mechanism of Eliminating Fluorescence

Concentrated nitric acid in the reaction system decomposed nitrogen
dioxide, with which the nitration reaction of unsaturated bond on the
vegetable oil molecular chain occurred at high temperature [22]. The
mechanism of reaction was proposed (see Figure 1).

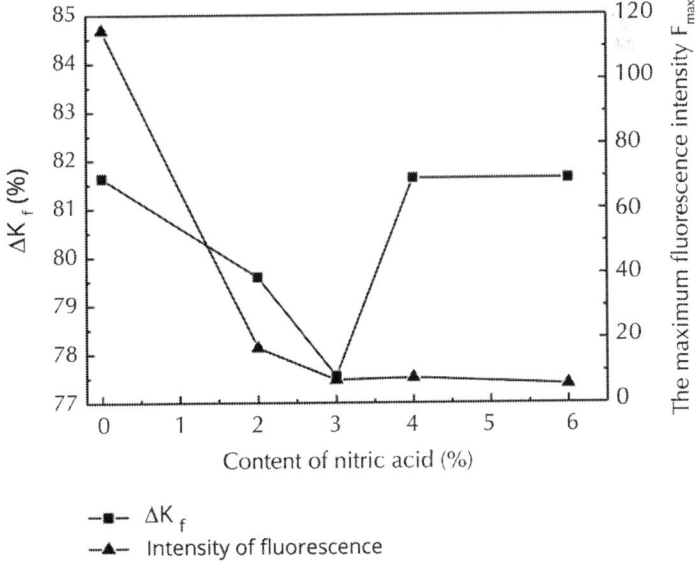

Figure 1

According to the theory of fluorescence spectroscopy, the main functional groups of fluorescence are double bond, conjugated double bonds, benzene, and fused ring structure with ϖ electron, wherein the benzene and fused ring structure have strong fluorescence emission property [23]. Therefore, after the oxidation of double bonds on the molecular chain of vegetable oil, the amount of ϖ bond decreases and the fluorescence intensity of lubricant is weakened.

With cheap and strong oxidation characteristics, the nitric acid was chosen as the fluorescent eliminating agent, which broke the ϖ bond of double bonds and reduced the fluorescence level, and then the low fluorescence lubricant was obtained.

The Optimization of Eliminating the Fluorescence

Effect of Nitric Acid Content on Properties of Low Fluorescence Lubricant

At a certain temperature, fixing the amount of lubricants, the liquid nitric acid was changed to react for 1 h, and the effect of nitric acid content on properties of low fluorescence lubricant was shown in Figure 2.

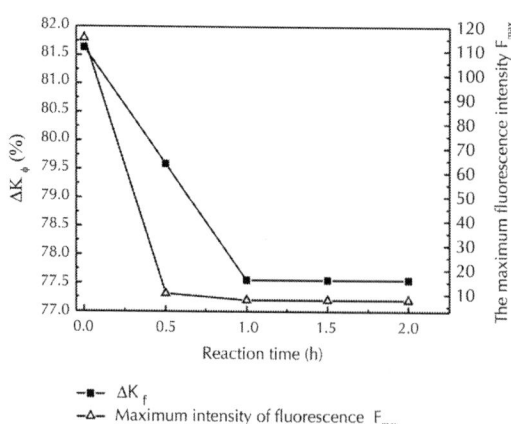

Figure 2: Effect of nitric acid content on properties of low fluorescence lubricant.

It can be seen from Figure 2 that the maximum fluorescence intensity of low fluorescence lubricant decreases gradually with the increase of nitric acid content, while the ΔK_f decreases first and then increases. When the dosage of nitric acid is 4% (wt/wt), the fluorescence properties showed the lowest maximum fluorescence intensity (F_{max} 7.725) and high value of ΔK_f (81.63%). So 4% of nitric acid content is chosen.

Effect of Reaction Time on Performance of Low Fluorescence Lubricant

At high temperature, the effect of reaction time on properties of low fluorescence lubricant was investigated after adding 4% of liquid nitric acid.

It can be seen in Figure 3 that the performance of low fluorescence lubricant presents regular changes with time changing. The value of reduced to remain changeless after 77.55%, while the maximum fluorescence intensity decreased in larger in half an hour and then changed little after one hour. As the reaction time increased, the unsaturated bonds were oxidized, then a part of the molecular chain was broken into small molecular compounds, leading to a reduction of the lubrication performance, and the fluorescence of lubricant intensity decreased meanwhile. In consideration of the fact that the value of K_f is 77.55% after 1 hour, the reaction time for eliminating fluorescence was 1 hour.

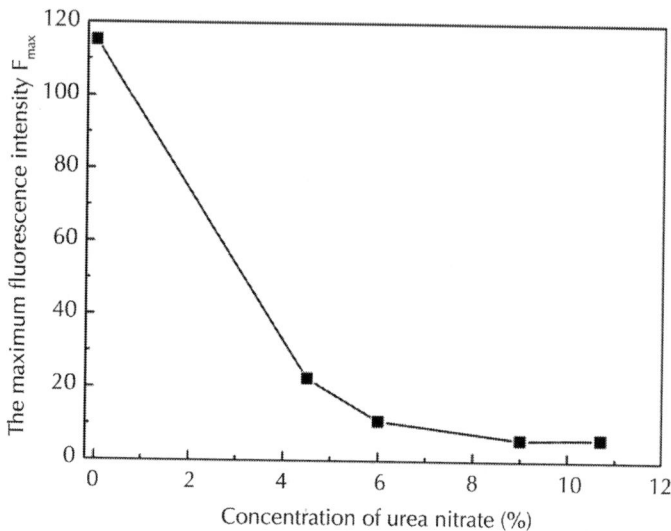

Figure 3: Effect of reaction time on performance of low fluorescence lubricant.

Influence of Urea Nitrate Amount on the Maximum Fluorescence Intensity of Low Fluorescing Lubricant

There is danger of using nitric acid liquid operating at high temperature; hence, it is necessary to use urea nitrate for instead. At the same conditions, the effect of urea nitrate content on performance of low fluorescence lubricant was also investigated at fixed reaction time (1 hour).

As shown in Figure 4, the value of the maximum fluorescence intensity reduced with the urea nitrate content increasing. The intensity value decreased to 5.871 when the amount of urea nitrate was increased to 9%. The low fluorescence lubricant produced by urea nitrate is with high freezing point and obvious particle, and it is easy to adhere to beaker. However, that produced by liquid nitrate does not have these disadvantages. Even so, the liquid urea nitrate is easy to store and use, and is of low cost as well. Whether using urea nitrate or liquid nitric acid as the fluorescent eliminating agent depends on the manufacturer.

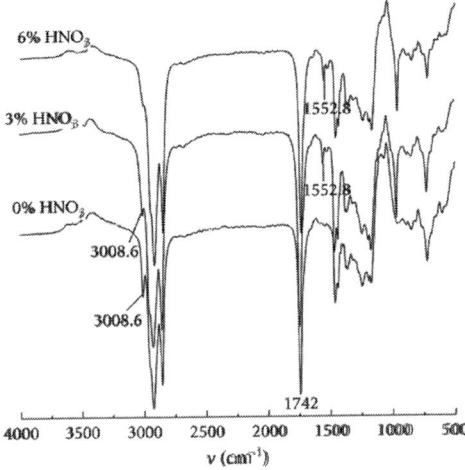

Figure 4: Effect of urea nitrate content on the maximum fluorescence intensity of low fluorescing lubricant.

The Effect of Eliminating Fluorescence Reaction on the Structure of Lubricant

The low fluorescence lubricants prepared with different dosages of nitric acid (reaction time 1 h) were characterized by FTIR, and the results were demonstrated in Figure 5.

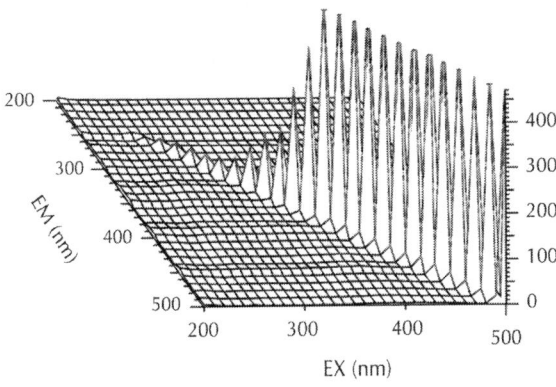

Figure 5: FTIR spectra of low fluorescing lubricant with different amount of nitric acid.

It can be seen that characteristic peak at 1742 cm⁻¹ is related to stretching vibration of carbonyl (C=O) for ester group. The peak at 3008.6 cm⁻¹ was assigned to =C–H and the 1552 cm⁻¹ peak was related to stretching vibration of –C–NO$_2$. The peak at 3008.6 cm⁻¹ was obvious without nitric acid, which decreased with the increase of nitric acid content, and the stretching vibration of –NO$_2$ at 1552 cm⁻¹ occurred simultaneously. When the nitrate concentration is 6%, the peak at 3008.6 cm⁻¹ is not obvious. The variation of peaks in the spectra illustrated that the nitration reaction took place with C=C bond broken and C–NO$_2$ formation.

Effect of Elimination of Fluorescent Reaction on the Fluorescence Characteristics of Lubricant

Table 1 demonstrated the fluorescence characteristics of lubricant before and after elimination of fluorescence, which showed that the maximum fluorescence intensity, the corresponding optimal excitation and emission wavelength are changed.

Table 1: Fluorescence characteristics of lubricant before and after elimination of fluorescence

Lubricant	The maximum fluorescence intensity Fmax	Corresponding EX (nm)	Corresponding EM (nm)
E167	7.725	258	332.0
Ordinary lubricant	115.2	300	385.6

Figures 6 and 7 are 3D fluorescence spectra of E167 and ordinary lubricant, respectively. Comparing Figures6 and 7, the ordinary lubricant shows fluorescence peak curve in three-dimensional map without nitrate. However, the fluorescence intensity peak of E167 was not presented on the map after eliminating fluorescence. The results illustrated that the low fluorescing lubricant E167 prepared by reacting 4% liquid nitric acid for 1 hour has low intensity of fluorescence. The strong peaks on the diagonal were produced by the instrument itself, which cannot be avoided when the scan ranges of both EX and MX were set at 200–500 nm.

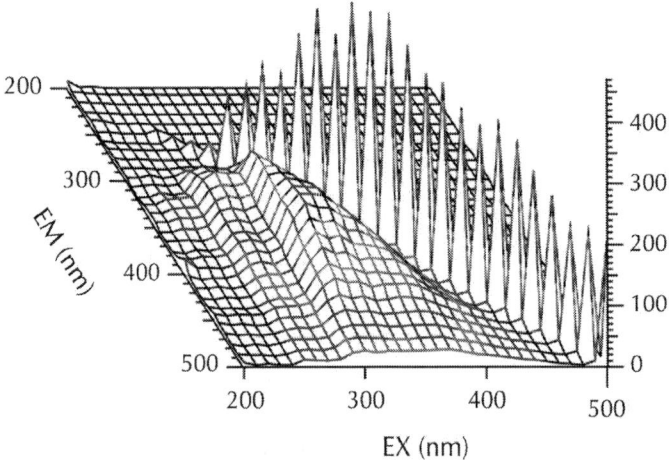

Figure 6: 3D fluorescence spectra of E167.

$$4HNO_3 \longrightarrow 2NO_2 + O_2 + 2H_2O$$

R−CH≡CH−CH₂−COO-CH₂
R−CH≡CH−CH₂−COO-CH + 2NO₂ $\xrightarrow{\text{T}}$
R−CH₂−CH₂−CH₂−COO-CH₂

RHC−HC−CH₂−COO-CH₂
R−CH≡CH−CH₂−COO-CH
R−CH₂-CH₂−CH₂-COO-CH₂

Figure 7: 3D fluorescence spectra of ordinary lubricant.

The Comprehensive Performance Evaluation of E167

Rheological Properties

Adding E167 into fresh water based mud and brine based mud, respectively, the rheological properties of E167 in 5.2% bentonite fresh water based mud (Weifang, Shandong) and 4% brine based mud systems were measured and the results were shown in Tables 2 and 3.

Table 2: Rheology variation of fresh water mud before and after adding E167

Experimental condition	Content of E167 (%)	AV (mPa·s)	PV (mPa·s)	YP (Pa)	FLAPI (mL)
Room temperature	0	10.5	6	4.3	19.2
	0.5	11.5	6	5.7	18.4
	1.0	12.5	6	6.2	16.8
150°C × 16h	0	17.0	8	9.1	26.0
	0.5	17.7	8	9.8	24.0
	1.0	18.0	8	10.1	24.0
180°C × 16h	0	18.0	7	5.1	24.0
	0.5	19.0	8	10.5	26.0
	1.0	17.3	8	7.4	24.0

Table 3: Rheology variation of brine mud before and after adding E167

Experimental condition	Content of E167 (%)	AV (mPa·s)	PV (mPa·s)	YP (Pa)	FLAPI (mL)
Room temperature	0	4.5	2	2.4	66
	0.5	4.8	3	2.2	66
	1.0	5.0	3	2.4	68
150°C × 16h	0	3.5	3	1.0	104
	0.5	3.7	3	1.2	96
	1.0	4.0	3	1.4	86
180°C × 16h	0	3.7	2	1.0	112
	0.5	4.0	2	2.4	106
	1.0	4.3	3	1.2	94

It can be seen from Table 2, in fresh water based mud, that both the apparent viscosity and yield point increased slightly, and the filter loss of API decreased mildly after adding into E167.

As shown in Table 3, in brine based mud, the same regularity was obtained: both the apparent viscosity and yield point increased slightly, and the filter loss of API decreased, meaning that the addition of E167 had no impact on rheological properties of the drilling fluid.

Foamability

The foaming performance at room temperature and thermal ageing at high temperature were evaluated after adding the E167 into fresh water based mud system. The methods of determining thermal ageing foaming rate are as follows: add 2.00 g sample into 400 mL fresh water based mud and put it into the aging tank after stirring 5 min. The mixture was thermal aged for 16 h at setting temperature, mixing with glass rod and then taking 300 mL mixture which was stirred for 5 min at 10000 r/min speed. After tiring, put the 300 mL mixture into 500 mL cylinder in 10 s, read the total volume at 30 s, and then calculate the foaming rate according to the formula.

Table 4 showed that, at the dosage of 1%, the E167 had weak foamability after thermal ageing for 16 h at 120°C, while the foaming was strong with the same time of hot rolling at 150°C. That is because the fat was hydrolyzed to form fatty acid at high temperature and then reacted with bases to form anionic surfactant sodium carboxylate, causing the enhancement of foaming capacity.

Table 4: Foaming rate of E167 at different conditions

Sample	Experimental condition	Foaming rate (%)
Distilled water + 0.5% X-E167	Room temperature	3.3
Fresh water mud + 1.0% X-E167	Room temperature	6.7
Fresh water mud + 1.0% X-E167	Thermal ageing 120°C × 16 h	5.0
Fresh water mud + 1.0% X-E167	Thermal ageing 150°C × 16 h	43.3

Fresh water mud is mixed with 3.7% bentonite and distilled water (Chifeng Tianyu), pH = 9, AV = 8-9 mPa·s.

Lubricity

Adding E167 into 5.2% bentonite fresh water based mud (Weifang, Shandong) and 4% brine based mud successively, the lubricity and salt resistance were investigated as shown in Table 5.

Table 5: The evaluation of lubricity for E167

Type of mud	Content of E167 (%)	ΔK_f (%)	Δf (%)	Testing condition
Fresh water mud	0.5	77.55	79.14	At room temperature
	1.0	77.55	76.82	At room temperature
	1.0	71.33	70.27	Thermal ageing 180°C × 16h and then at room temperature
Brine mud	0.5	75.00	62.43	At room temperature
	1.0	77.08	66.60	At room temperature
	1.0	73.55	64.30	Thermal ageing 180°C × 16h and then at room temperature

It can be seen from the table that the E167 has good lubricant performance. In case of changing conditions, such as thermal aged at 180∘ C for 16 h and even in the 4% brine based mud, the values of K_f and f changed a little, which means that the E167 with good lubricity has relatively good temperature resistance and salt resistance.

Comparison of Properties with Other Low Fluorescence Lubricants

We selected two kinds of oil based low fluorescence lubricant with comprehensive performance on scene as a contrast sample, which were from Shengli Oilfield (S-WD) and Tarim Oilfield (T-YS), respectively. The properties of lubrication, fluorescence, and foaming rate were determined as shown in Table 6.

Table 6: Comparison of three kinds of oil based properties of low fluorescence lubricant

Properties	Methods and technical indicators	Low fluorescence lubricant		
		E167	S-WD	T-YS
Lubricity	ΔKf (%)	77.55	75.51	77.55
	Δf (%)	76.82	77.02	68.07
Fluorescence performance		7.08	10.50	10.98
	Corresponding of Fmax EX (nm)	258	274	376
	Corresponding of Fmax EM (nm)	332	325	444
Foaming rate (%)	Mud + 1% X-E167, at room temperature	5.0	8.3	10.0
	Mud + 1% X-E167, after thermal ageing 120°C × 16 h	6.7	13.0	31.7
	Mud + 1% X-E167, after thermal ageing 150°C × 16 h	43.3	48.0	45.0

Compared with commercially available excellent products, the low fluorescence lubricant E167 prepared in this work has outstanding properties. Moreover, the comprehensive foaming performance after thermal ageing was superior to the contrast sample.

CONCLUSIONS

In summary, the low fluorescence lubricant used for drilling fluid was prepared, taking animal and vegetable oil as raw material. The properties were studied by many methods. It is revealed from the experimental results that, with the addition of concentrated nitric acid, the modified lubricant was qualified in low fluorescence and comprehensive good lubricating performance. In view of the influence of temperature, the E167 exhibited good lubricity after thermal ageing at 180°C for 16 hours showing properties of high-temperature resistance. The foaming rate of E167 was lower when compared with similar products although which exhibited serious foaming phenomenon.

REFERENCES

1.   D.-J. Li, "Status and advances of drilling fluid lubricants," Petroleum Drilling Techniques, vol. 26, no. 2, pp. 35–38, 1998.

2.   R. Caenn, H. C. H. Darley, and G. R. Gray, Composition and Properties of Drilling and Completion Fluids, Gulf Professional Publishing, 6th edition, 2011.

3.   G.-J. Lv, "Low-fluorescence anti-blocking lubricant for drilling fluid and production method thereof," CN, 101717621 A, 2010-6-2.

4.   G.-S. Feng, C.-S. Li, H. Tang, and W.-J. Wang, "Compound type vegetable oil lubricating agent for drilling fluid and preparation method thereof," CN, 101760186 A, 2010-6-30.

5.   A. D. Patel, E. Stamatakis, S. Young, et al., "High performance water based drilling fluid," US: 2008/0009422 A1, 2008-1-10.

6.   D. Knox and P. Jiang, "Drilling further with water-based fluids—selecting the right lubricant," inProceedings of the SPE International Symposium on Oilfield Chemistry, pp. 9–15, The Woodlands, Tex, USA, February 2005.

7.   H. Wagner, R. Luther, and T. Mang, "Lubricant base fluids based on renewable raw materials: their catalytic manufacture and modification," Applied Catalysis A, vol. 221, no. 1-2, pp. 429–442, 2001. · ·

8.   N. J. Fox and G. W. Stachowiak, "Vegetable oil-based lubricants-A review of oxidation," Tribology International, vol. 40, no. 7, pp. 1035–1046, 2007. · ·

9.   P. Mousavi, D. Wang, C. S. Grant, W. Oxenham, and P. J. Hauser, "Measuring thermal degradation of a polyol ester lubricant in liquid phase," Industrial and Engineering Chemistry Research, vol. 44, no. 15, pp. 5455–5465, 2005. · ·

10. S. Z. Erhan, B. K. Sharma, Z. Liu, and A. Adhvaryu, "Lubricant base stock potential of chemically modified vegetable oils," Journal of Agricultural and Food Chemistry, vol. 56, no. 19, pp. 8919–8925, 2008. · ·

11. B.-R. Höhn, K. Michaelis, and R. Döbereiner, "Load carrying capacity properties of fast biodegradable gear lubricants©," Lubrication Engineering, vol. 55, no. 11, pp. 15–38, 1999. ·

12. B. Krzan and J. Vizintin, "Tribolngical properties of an environmentally adopted universal tractor transmission oil based on vegetable oil," Tribology International, no. 36, pp. 826–832, 2003.

13. J.-H. Liu and A.-Y. Qiu, "New progress in hydrogenation technology of vegetable oil(I)," China Oils and Fats, vol. 28, no. 8, pp. 13–17, 2003.

14. S.-Y. Song, "Drilling fluid lubricating additive and its preparing method," CN, 1743404A, 2006-3-8.

15. H. Chun, W.-L. Xie, and W. Guo, "Study on synthesis and characterization of triethanolamine dilaurate," Cereals & Oils, no. 7, pp. 11–13, 2008.

16. X. Wu, X. Zhang, S. Yang, H. Chen, and D. Wang, "Study of epoxidized rapeseed oil used as a potential biodegradable lubricant," Journal of the American Oil Chemists› Society, vol. 77, no. 5, pp. 561–563, 2000. ·

17. A.-J. Liu, "Understanding and pondering over fluorescence problem of drilling fluid additives,"Drilling Fluid and Completion Fluid, vol. 20, no. 2, pp. 9–12, 2003.

18. A. D. Patel, "Non-fluorescing oil-based drilling fluid," US, 5869433, 1999-2-9.

19. Q/SY, 1088-2007, "Specifications of liquid lubricants used in drilling fluids," China National Petroleum Corporation. 12, 2006.

20. GB-T, 16783. 1-2006, "Petroleum and natural gas industries-Field testing of drilling fluids-Part 1: water-based fluids," China National Standard. 12, 2006.

21. Q/SY TZ, 0022-2000, "Specifications and analytical methods for drilling fluid lubricants," Tarim Oilfield of China National Petroleum Corporation. 12, 2000.

22. D. Hace, V. Kovacevic, and D. Pajc-Liplin, "Thermally stimulated oxidative degradation of high impact polystyrene with nitric acid," Polymer Engineering and Science, vol. 36, no. 8, pp. 1140–1151, 1996. ·

23. Y.-Q. Liu, Modern Instrumental Analysis, Higher Education Press, Beijing, China, 2006.

Chapter 3

Chemical Effect on Wellbore Instability of Nahr Umr Shale

Baohua Yu[1], Chuanliang Yan[1], and Zhen Nie[2]

[1]State Key Laboratory of Petroleum Resource and Prospecting, China University of Petroleum, Beijing 102249, China

[2]Exploration & Production Research Institute, CNPC, Beijing 100011, China

ABSTRACT

Wellbore instability is one of the major problems that hamper the drilling speed in Halfaya Oilfield. Comprehensive analysis of geological and engineering data indicates that Halfaya Oilfield features fractured shale in the Nahr Umr Formation. Complex accidents such as wellbore collapse and sticking emerged frequently in this formation. Tests and theoretical analysis revealed that wellbore instability in the Halfaya Oilfield was influenced by chemical effect of fractured shale and the formation water with high ionic concentration. The influence of three types of drilling fluids on the rock mechanical properties of Nahr Umr Shale is tested, and time-dependent collapse pressure is calculated. Finally, we put forward engineering countermeasures for safety drilling in Halfaya Oilfield and point out that increasing the ionic concentration and improving the sealing capacity of the drilling fluid are the way to keep the wellbore stable.

INTRODUCTION

The Nahr Umr Shale Formation is found throughout the southern part of the Arabian Gulf and forms the cap rock to many major reservoirs in the region [1]. Major wellbore instability problems when drilling through this shale formation have often arisen not only in new wells but also in reentry wells, especially with the rise of water-based mud and stricter environmental control, making wellbore stability in this shale an extremely challenging operation for drilling/mud engineers.

The Halfaya Oilfield is in the south of Missan province in Iraq, which is 400 km south east of Baghdad, the capital of Iraq. Three horizontal wells in the Nahr Umr Formation of Halfaya Oilfield had been drilled. But, two wells of the three horizontal wells have sidetracking due to sticking, only one horizontal well is drilled successfully, so it illustrates the big effects of wellbore instability problems on the directional drilling in this oilfield.

Geological Character

The Halfaya Oilfield is located on the Arabian shelf, which is adjacent to the Zagros tectonic zone. The influence of Zagros tectonic movement is the extrusion to the Arabian shelf by the European plate (NNE-SSW). The propagation of the stress wave leads to a series of anticlines in the Arab shell. This extrusion stopped in Middle Miocene. The geological structure is a low dip anticline, in which the long axis is nearly perpendicular to the Zagros extrusion stress field [2–4]. The structure is above the Arabian shelf and is far away from the Zagros fault control zone, but the structure is still affected by Zagros tectonic movement, which makes the in situ stress complicated.

There is no large fault which could be recognized by seismic data. The anticline structure is also very smooth. The results show that the extrusion stress by the Zagros tectonic movement is not very strong, and the extrusion stress does not produce strong in situ deformation and destruction.

The lithologic characters in the Halfaya Oilfield from the top to the bottom is, respectively, the Tertiary Upper Fars Group, mainly sandy mudstone, about 1300 m thick; the Lower Fars Group, mainly anhydrite, salt rock, and shale deposit, about 500 m thick, being the regional cap rock; the Tertiary Kirkuk Group which is mainly sandstone and mudstone, about 300 m; from the Tertiary Jaddala group to the Nahr Umr group, mainly carbonatite and interlayers of thin marl, sandstone and shale.

DRILLING PROBLEMS

Three horizontal wells, N001H, N006H, and N002H, in the Nahr Umr Formation of Halfaya Oilfield had been drilled. The well distributions are both located in the structural long axis direction. The complicated drilling problems of these three wells are as follows.

When the first horizontal well (N001ST well) encountered the Nahr Umr layer, there were two sidetracking operations. The first sidetracking happened at 3941.26 m, the SLB screw stuck in the highly deviated interval and the directional tool dropped in the well. The fishing failed, which led to sidetracking. The second sidetracking happened at 4091.21 m, the Nahr Umr Shale collapsed and this led to sticking at 4087 m the treatment measures was ineffective and a sidetracking operation happened.

The second horizontal well (N006ST well) used the organic salt drilling fluid which has a strong inhibition. When drilling 3964 m, the Nahr Umr Shale collapsed and the treatment measure for the sticking failed, so sidetracking happened at 3800 m using the vertical well completion.

The third horizontal well N002H used saturated salt water drilling, but there were many sticks between 3660 m and 3895 m in the Nahr Umr Formation, and there were cavings at the shaking screen.

The two wells of the three horizontal wells have sidetracking due to sticking; only one horizontal well was drilled successfully, so it illustrates the big effects of wellbore instability problems in the oilfield. The wellbore instability has serious impact for oilfield drilling, and it restricts the exploration and development in the oil field. As the development wells are generally directional wells or horizontal wells, and the wellbore instability risk is great, wellbore stability analysis is needed in the Halfaya Oilfield.

WELLBORE INSTABILITY MECHANISM

Character of the Instable Shale

Figure 1 shows the logging data of the Nahr Umr Formation of N004 well. The GR logging shows that the formations are mainly sandstone and shale. The caliper logging shows that there are both stable and instable intervals. Compared to the GR logging data, the lithology of collapsing interval is shale and the sandstone interval is stable. According to the interval transit time logging data, the interval transit time of the Nahr Umr Shale is obviously higher than the adjacent sandstone interval. The density logging data of the shale are obviously lower than the adjacent sandstone interval. The reasons for this phenomenon are the rich internal microfractures, drilling fluid, and the filtrate seepage. This can be seen from the photo of the Nahr Umr Shale (Figure 2). In addition, the shape of cavings of the Nahr Umr Shale indicates that the shale is enriched in fractures (Figure 3).

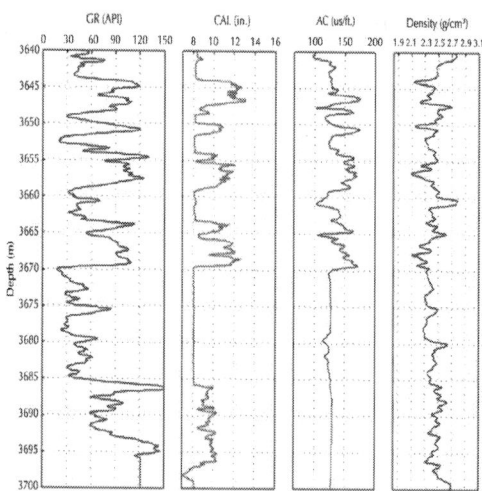

Figure 1: The comparison of the logging data in Nahr Umr Formation. GR: natural gamma logging; CAL: caliper logging; AC: acoustic transit time logging.

Figure 2: The core of Nahr Umr Shale.

Figure 3: The shale cavings of the Nahr Umr Shale of N006H Well.

The main reason for wellbore instability in hard brittle shale with lots of fractures is as follows [5–11]. If the sealing capacity of the drilling fluid is not enough or the ionic concentration is not enough to balance the formation water ionic concentration, the drilling fluid and the filtrate would flow into the microfractures under the driving power from the fluid column pressure difference of drilling fluid and the ionic concentration difference. This would lead the friction coefficients of the fracture plane to decrease, the effective stresses around the wellbore to decrease, the formation around the wellbore to become loose, and the support of the drilling fluid column to the wellbore wall decrease. Thus the formation fluid will flow into the wellbore. During the reaming and back reaming, the disturbance of the rigs to the loose formation will lead to wellbore instability.

Wellbore Instability Analysis

In order to solve the wellbore instability of the fractured and hard-brittle shale in Nahr Umr Formation, the drilling fluid property and engineering countermeasures should be taken into account. We analyze the wellbore instability reasons combined with the drilling engineering measures of three horizontal wells in the Nahr Umr Formation.

When using rational engineering measures, the drilling fluid property decides the wellbore stability during drilling to a large extent. Table 1 show the drilling fluid property used in these three horizontal wells. These three wells used three different types of drilling fluid. The following can be concluded from the drilling fluid property parameters in the table.

Table 1: The drilling fluid properties of three horizontal wells of Nahr Umr Formation

	Well no.			
	N001H well-Hole 1	N001H well-Hole 2	N002H well	N006H well
Mud type	KCL-polymer	**KCL-polymer**	**Salt saturated**	**BH-WEI**
Density (g/cm³)	1.25	1.25	1.28	1.28
Viscosity (s)	51	53	78	65
Plastic viscosity (cp)	26	27	41	39
Y.P (lb/100 ft²)	24	26	31	29
Gel strength 10"/10' (lb/100 ft²)	5/8	5/14	7/9	5/7
API filtrate (mL)	3.2	3.4	3.0	3
Mud cake (mm)	0.3	0.3	0.3	0.5
PH	9.5	9	9	8.5
Solid (%)	13	11	13	17
Sand (%)	0.3	0.3	0.2	0.3
Bentonite content (g/L)	27	26	38	
Potassium (mg/L)			27000	
Chloride (mg/L)			55000	11520
Ca⁺ (mg/L)			200	

- Based on the mud rheological parameters, for the formation with good completeness, the rheological parameters of these three wells are similar and could meet the engineering requirement. But, for the fractured shale formation, the rheological parameters of these three wells are different. Compared to the other two wells, the drilling fluid of N001H well has a low viscosity, which is bad for carrying the cuttings and cavings. In addition, the drilling fluid with low viscosity will easily flow into the formation under the pressure difference. Therefore, the rheological parameters of drilling fluid of N002H well benefit wellbore stability. Usually increasing the drilling fluid viscosity is of benefit for fracture formation.

- Based on the drilling fluid filter loss, the filter losses of these three wells are similar. Because the filter loss is measured by the experimental instrument in the laboratory, the results cannot reflect the real formation situation and it is only a reference index.

- Based on the drilling fluid ionic concentration, although there are not ionic concentration parameters of N001H well drilling fluid in the daily drilling report, according to the drilling fluid description provided by the drilling fluid service provider, the drilling fluid ionic concentrations of this well could indicate that the ionic concentration of KCL Polymer drilling fluid used in the N001H well is between the concentrations of the N002H well and N006H well; the ionic concentration of the N002H well is the highest the ionic concentration of the N006H well is the lowest. When the hole is opened, the ionic concentration difference of the drilling fluid and formation water is the main driving force that drives the free water in the drilling fluid into the formation. Commonly, the high ionic concentration of drilling fluid is of benefit to prevent the free water in the drilling fluid from flowing into the formation. If the free water in the drilling fluid flows into the formation, the formation will be hydrated, and the formation strength will be decreased, so as to lead to wellbore periodic collapsing. Table 2 shows the formation water property of Halfaya Oilfield. The results show that the formation water has an extremely high ionic concentration, which needs a high ionic concentration for drilling fluid to balance it.

Table 2: The formation fluid properties of Halfaya oilfield

	Unit	Nahr Umr
Water type		$CaCl_2$
PH		6.3
Specific gravity (15.56°C)	sg	1.121
Resistivity (25°C)	ohm·m	0.068
Total salinity	ppm	166661
Total hardness	mg/L	16562
Na^+	mg/L	60015
Ca^{2+}	mg/L	8681
Mg^{2+}	mg/L	993
Fe^{2+}	mg/L	74
Ba^{2+}	mg/L	1
K^+	mg/L	716
Sr^{2+}	mg/L	356
Cl^-	mg/L	107098
SO_4^{2-}	mg/L	874
HCO_3^-	mg/L	7263
CO_3^{2-}	mg/L	0
OH^-	mg/L	0

Shale Hydration

According to the formation character, the Nahr Umr Shale is abundant in microfractures; the formation is broken, and the drilling fluid can easily flow into the micro fracture plane, which leads to the change of formation strength. In order to prevent wellbore instability, the drilling fluid property should be improved. The influence of drilling fluid on the wellbore stability is analyzed from the mineral composition, the drilling fluid consistency, and the influence of the drilling fluid on formation strength.

Tables 3 and 4 illustrate the minerals and clay minerals composition and content of the Nahr Umr Shale, respectively. The test results in the tables show that the Nahr Umr Shale mainly consists of quartz and clay, especially quartz, which exceeds 48.5%. For shale Formation, the higher the quartz, the higher the brittleness; at the same time, the content of the clay minerals of the shale belongs to medium and little high level. The clay minerals mainly consist of illite/smectite and kaolinite, and the content of smectite is low. The type of the clay mineral indicates that the shale is very brittle. In addition, the kaolinite is a stable clay mineral, and the hydration of the illite/smectite is also feeble. The type and content of the clay minerals both indicate that Nahr Umr Shale Formation is a hard and brittle formation which is hard to hydrate.

Table 3: Mineral composition and content of the Nahr Umr Shale

Depth	Mineral content (%)								
	Quartz	Potassium feldspar	Soda feldspar	Anorthose	Calcite	Dolomite	Iron pyrite	Hematite	TCCM
3645.10	51.7	0.8		0.2				2.7	44.6
3649.83	60.8	1.2			0.4		4.7		32.9
3666.00	48.5	1.9		0.3	1.4		4.5		43.4

Table 4: Clay mineral composition and content of the Nahr Umr Shale

Depth	Clay mineral content (%)						Interbed ratio (% S)		
	S	I/S	It	Kao	C	C/S	I/S	C/S
3645.10		34	7	48	11		14	
3649.83		33	3	40	24		11	
3666.00		44	7	49			21	

According to research experience, if the drilling fluid inhibition is good enough, a formation like the Nahr Umr Shale is impossible to hydrate without expansion collapsing. Therefore, we evaluate the rejection capacity of three drilling fluid systems which are used in the Halfaya Oilfield. The three drilling fluids are organic salt drilling fluid, Gel-polymer drilling fluid, and KCl-polymer drilling fluid.

The densities of these three types of drilling fluid are all 1.33 g/cm^3; then we measured the cuttings recovery (The 40 g cuttings with 3.2~2.0 mm diameter injected to the 350 mL fluid. Roll 16 h in a set temperature. Then filter the cuttings through a sieve. Dry and weigh cuttings to calculate the cuttings recovery.) and swelling ratio of these three drilling fluids The results are shown in Table 5. The results show that the cuttings recoveries of these three types of drilling fluid are both higher than 95% for the Nahr Umr Shale; although the swelling ratios are different. These results show that the inhibitive capacity of the drilling fluid is good [12]; on the other hand, the results indicate that the formation hydration is feeble. The inhibitive capacity of the drilling fluid is not the main reason for the wellbore instability of the Nahr Umr Shale.

Table 5: Swelling ratio and recovery of the Nahr Umr Shale

	Organic salt	KCL-polymer	Gel-polymer
Recovery Rate (%)	95	96	97
Swelling Ratio (%)	24	36	22

In order to analyze the influence of the drilling fluid on the wellbore stability of the Nahr Umr Shale, experimental studies were carried on the influence of drilling fluid on the rock mechanical property. We tested the shale strength of Nahr Umr Shale after immersing it in different kinds of drilling fluids. Table 6shows the uniaxial compressive strength (UCS, MPa) results from the test. Figure 4 illustrates the comparison of the strength variation rule versus the time after immersing in different kinds of drilling fluid.

Table 6: Experimental results of shale UCS after immersing in drilling fluid

Drilling fluid type	Organic salt	KCL-polymer	Gel-polymer
UCS without immersing (MPa)	48.62	51.09	47.22
UCS with immersing of 24 h (MPa)	40.16	44.8	44.8
UCS with immersing of 48 h (MPa)	37.81	41.41	43.02
UCS with immersing of 72 h (MPa)	35.64	39.82	41.69
UCS with immersing of 96 h (MPa)	34.96	39	40.33

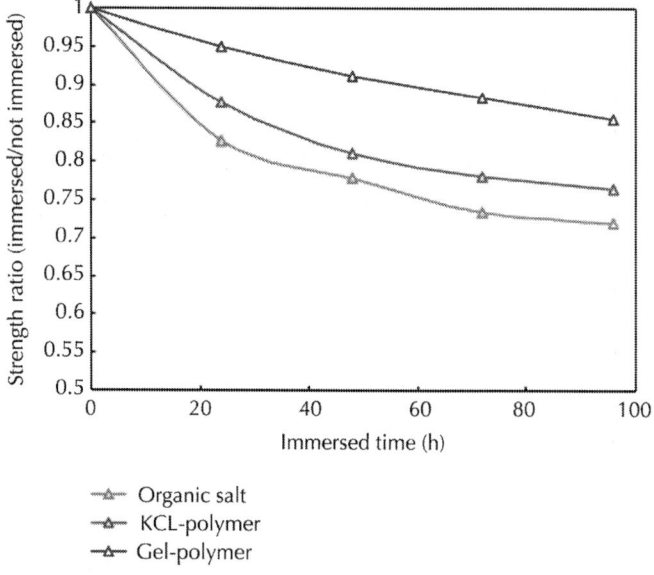

-△- Organic salt
-△- KCL-polymer
-△- Gel-polymer

Figure 4: Comparison of the shale strength decrease after immersing.

Figure 4 shows that the shale UCS decreases greatly after immersing it in the organic salt drilling fluid, the next is the KCL-polymer drilling fluid; the strength in the Gel-polymer drilling fluid changed a little. Therefore, the Gel-polymer drilling fluid benefits the wellbore stability of the Nahr Umr Shale.

Under the drive force of the ionic concentration difference, the free water in the drilling fluid which flows into the formation would decrease the rock strength, which is the main reason for the collapse in Nahr Umr Shale; in addition, as the formation is extremely hard and brittle and the fractures are rich internally, if the drilling fluid sealing capacity is not good enough, the drilling fluid and filtrate would flow into the rock along the microfracture under the difference of the drilling fluid column pressure and pore pressure, so as to weaken the formation strength and lead to wellbore collapse. So, the increasing of the ionic concentration of the drilling fluid and enhancing the drilling fluid sealing capacity is the key to the wellbore stability of the Nahr Umr Shale.

TIME-DEPENDENT COLLAPSE PRESSURE

According to mechanical concepts, the main reason for borehole collapse is caused by shear failure for the reason that stresses loaded on rock around the borehole exceed the rock strength, as a result of lower mud column pressure. Now, brittle formation collapse will generate and the borehole will enlarge; for plastic formation, plastic deformation is will generated and borehole shrinkage will be encountered.

Generally, borehole collapse takes place in the minimum horizontal stress direction, $\theta = \pi/2$ or $\theta = 3\pi/2$ [12]; the borehole stress on minimum horizontal stress direction [13–20] is as follows:

$$\sigma_r = P - \delta\phi\left(P - P_p\right),$$

$$\sigma_\theta = 3\sigma_H - \sigma_h - P + \delta\left[\frac{\alpha(1 - 2\nu)}{1 - \nu} - \phi\right]\left(P - P_p\right),$$

$$\sigma_z = \sigma_\nu + 2\nu\left(\sigma_H - \sigma_h\right) + \delta\left[\frac{\alpha(1 - 2\nu)}{1 - \nu} - \phi\right]\left(P - P_p\right). \tag{1}$$

Assume that safe coefficient FS [21] is as follows:

$$FS = \frac{\sigma_n tg\varphi + C}{\tau}.$$

(2)

And let

$$M = 1 + (FS - 1)\cos^2\varphi.$$

(3)

Replace normal stress σ_n as principal stress σ_1 and σ_3:

$$\sigma_n = \frac{\sigma_1 + \sigma_3}{2} - \frac{\sigma_1 - \sigma_3}{2}\sin\varphi - \alpha P_p.$$

(4)

Rewrite Mohr-Coulomb criterion [21]:

$$M(\sigma_1 - \sigma_3) - \sin\varphi(\sigma_1 + \sigma_3 - 2\alpha P_p) - 2C\cos\varphi = 0.$$

(5)

Based on different borehole stress conditions, borehole collapsing pressure expresses as a different form. When the bearing condition is $\sigma_\theta > \sigma_z > \sigma_r$, maximum and minimum stress separately are $\sigma_1 = \sigma_\theta$, $\sigma_3 = \sigma_r$, and make $k = (\alpha(1-2v)/(1-v)) - \phi$; under mud penetrating borehole face, the borehole collapsing pressure model is

$$P_{cr} = \left(2C\cos\phi + (\sin\phi - M)(3\sigma_H - \sigma_h)\right)$$

$$+ \left[\delta Mk + \sin\phi(2\delta f - \delta k - 2k)\right]P_p\right)$$

$$\times (\delta k\sin\phi - M(2 + \delta k - 2\delta f))^{-1}.$$

(6)

For the hard-brittle shale of Nahr Umr Formation, according to the study results and experimental results, the influence of the drilling fluid immersion on the mechanical property mainly reflects in the decrease of the compressive strength as the immersing time increase.

Figure 5 illustrates the variation of collapse pressure versus the hole opening time for Nahr Umr Shale. The collapse pressure would increase as the formation strength decreases; the increasing speed decreases gradually. The increasing rate of Gel-polymer drilling fluid is the lowest; in a certain drilling fluid density it can keep the wellbore stability for the longest time. The increasing of the mud density could only keep the wellbore stability in limited time. If the property of the drilling fluid cannot be improved, increasing the mud density would force the drilling fluid to flow into the formation and make the wellbore unstable.

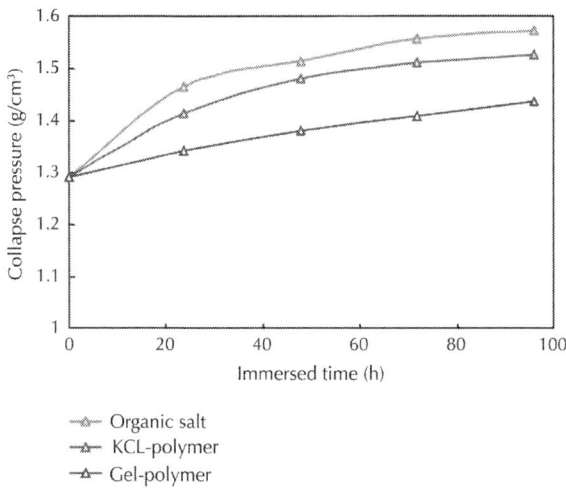

Figure 5: Time-dependent collapse pressure of Nahr Umr Shale.

COUNTERMEASURES DEALING WITH WELLBORE INSTABILITY

In order to prevent the wellbore instability of Nahr Umr Shale, we come up with the following drilling technology countermeasures and suggestions.

- Depending only on the drilling fluid density cannot solve the wellbore stability of the shale formation which is full of fractures [22, 23]. If the drilling density is too high, the pore pressure would increase and the effective stresses around the wellbore decrease, and this would cause a larger damaged scale. Decreasing the drilling fluid filter loss and improving the drilling fluid rheological property would benefit wellbore stability.

- Commonly, the larger the inclination, the more possible the wellbore instability. But for the laminar fracture formation, decreasing the angle of the wellbore axial line with the bedding normal direction is of benefit for the wellbore stability.
- The influences of the swabbing pressure and surge pressure should be taken into consideration when evaluating wellbore stability; the simplified bottom hole assembly (BHA) could prevent large swabbing pressure and surge pressure and then prevent sticking.
- The hydraulic jetting is not suitable, because the high pressure hydraulic jetting would produce water wedge effect in the progress of the drilling seepage. The big diameter jet or no-jet is welcomed.
- Avoiding the intense change of the dogleg or the well track so as to prevent big drill string acting force to the wellbore wall.
- Optimizing the hydraulic parameters so as to ensure the cuttings could be carried out of the wellbore timely. For some situations, wellbore collapse cannot be prevented, so carrying out the cuttings in a timely way could decrease the downhole complicated time. Increasing the drilling rate could decrease the exposed time of the shale formation, which is useful for the wellbore stability.
- The formation water has an extremely high ionic concentration, so keep a high ionic concentration for the drilling fluid to balance it.

CONCLUSIONS

Under the function of the ionic concentration difference, the free water in the drilling fluid which flows into the formation will decrease the rock compressive strength, which is the main reason for the collapse in Nahr Umr Shale; in addition, as the formation is extremely hard and brittle and the fractures are internally rich, if the drilling fluid sealing capacity is not good enough, the drilling fluid and filtrate will flow into the rock along the micro fracture surface under the difference of the drilling fluid column pressure and pore pressure, so as to weaken the formation strength and lead to wellbore collapse. So, increasing the ionic concentration of the drilling fluid to enhance the drilling fluid sealing capacity is the key point to the wellbore stability of the Nahr Umr Shale Formation.

The collapse pressure will increase as the formation strength decreases after drilling; the increasing speed decreases gradually. The increasing rate of Gel-polymer drilling fluid is the lowest; in a certain drilling fluid density it can keep the wellbore stable for the longest time. The increase of the mud density could only keep the wellbore stability for a limited time. Improving the property of the drilling fluid is the basis for keeping the wellbore stable.

ACKNOWLEDGMENTS

This work is financially supported by the Science Fund for Creative Research Groups of the National Natural Science Foundation of China (Grant no. 51221003) and National Oil and Gas Major Project of China (Grant no. 2011ZX05009-005).

REFERENCES

1. V. X. Nguyen, Y. N. Abousleiman, and S. K. Hoang, "Analyses of wellbore instability in drilling through chemically active fractured-rock formations," SPE Journal, vol. 14, no. 2, pp. 283–301, 2009. ·

2. Z. Nie, H. Liu, A. Liu, et al., "The large temperature difference, long column and narrow clearance cementing best practices in Halfaya oilfield," in SPE Asia Pacific Oil and Gas Conference and Exhibition, October 2012.

3. M. Zhang, Z. Tian, S. Xu, et al., "Research and application of BH-ATH (anti-three high) drilling fluid system," in IADC/SPE Asia Pacific Drilling Technology Conference and Exhibition, July 2012.

4. H. Rabia, Iraqi Oil Reserves: Opportunities for Production and Exploration, 2008.

5. B. S. Aadnoy, "Introduction to special issue on borehole stability," Journal of Petroleum Science and Engineering, vol. 38, no. 3-4, pp. 79–82, 2003. ·

6. B. S. Aadony, "Modeling of the stability of highly inclined boreholes in anisotropic rock formations,"SPE Drilling Engineering, vol. 3, no. 3, pp. 259–268, 1987.

7. F. J. Santarelli, C. Dardeau, and C. Zurdo, "Drilling through hinghly fractured formations: a problem, a medel, and a cure," in Proceedings of the SPE Annual Technical Conference and Exhibition, October 1992.

8. O. A. Helstrup, Z. Chen, and S. S. Rahman, "Time-dependent wellbore instability and ballooning in naturally fractured formations," Journal of Petroleum Science and Engineering, vol. 43, no. 1-2, pp. 113–128, 2004. · ·

9. R. Narayanasamy, D. Barr, and A. Milne, "Wellbore instability predictions within the cretaceous mudstones, clair field, West of Shetlands," in Offshore Europe, Paper SPE 124464, Aberdeen, UK, 2009.

10. B. H. Yu, Study on Borehole Unstable Mechanism of Layered Shale, China University of Petroleum, Beijing, China, 2006.

11. J. L. Yuan, J. G. Deng, Q. Tan, B. H. Yu, and X. C. Jin, "Borehole stability analysis of horizontal drilling in shale gas reservoirs," Rock Mechanics and Rock Engineering, vol. 46, no. 5, pp. 1157–1164, 2013.

12. J. N. Yan, Drilling Fluid Technology, China University of Petroleum Press, 2001.

13. E. Fjær, R. M. Holt, P. Horsrud, et al., Petroleum Related Rock Mechanics, Elsevier, 2nd edition, 2008.

14. J. S. Bell and D. I. Gough, "Northeast-Southwest compressive stress in Alberta evidence from oil wells," Earth and Planetary Science Letters, vol. 45, no. 2, pp. 475–482, 1979. ·

15. D. I. Gough and J. S. Bell, "Stress orientations from borehole wall fractures with examples from Colorado, East Texas, and Northern Canada," Canadian Journal of Earth Sciences, vol. 19, no. 7, pp. 1358–1370, 1982. ·

16. M. D. Zoback, D. Moos, L. Mastin, and R. N. Anderson, "Well bore breakouts and in situ stress,"Journal of Geophysical Research, vol. 90, no. 7, pp. 5523–5530, 1985. ·

17. S. H. Hickman, J. H. Healy, and M. D. Zoback, "In situ stress, natural fracture distribution, and borehole elongation in the

Auburn geothermal well, Auburn, New York," Journal of Geophysical Research, vol. 90, no. 7, pp. 5497–5512, 1985. ·

18. C. A. Barton, M. D. Zoback, and K. L. Burns, "In-situ stress orientation and magnitude at the Fenton Geothermal site, New Mexico, determined from wellbore breakouts," Geophysical Research Letters, vol. 15, no. 5, pp. 467–470, 1988. ·

19. B. C. Haimson and C. Chang, "True triaxial strength of the KTB amphibolite under borehole wall conditions and its use to estimate the maximum horizontal in situ stress," Journal of Geophysical Research, vol. 107, no. 10, pp. ETG 15-1–ETG 15-14, 2002. · ·

20. M. Chen, Y. Jin, and G. Q. Zhang, Petroleum Engineering Related Rock Mechanics, Science Press, Beijing, China, 2008.

21. J. G. Deng, Y. F. Cheng, M. Chen, and B. H. Yu, Wellbore Stability Evaluation Technique, Petroleum Industry Press, 2008.

22. V. Maury and C. Zurdo, "Drilling-induced lateral shifts along pre-existing fractures: a common cause of drilling problems," SPE Drilling & Completion, vol. 11, no. 1, pp. 17–24, 1996. ·

23. X. Chen and C. P. Tan, "The impact of mud infiltration on wellbore stability in fractured rock masses," in SPE/ISRM Rock Mechanics Conference, 78241, October 2002.

Chapter 4

A Comparative Study of Diesel Oil and Soybean Oil as Oil-Based Drilling Mud

Okorie E. Agwu, Anietie N. Okon,
and Francis D. Udoh

Department of Chemical & Petroleum Engineering, Faculty of Engineering, University of Uyo, PMB 1017, Uyo, Akwa Ibom State 52001, Nigeria

ABSTRACT

Oil-based mud (OBM) was formulated with soybean oil extracted from soybean using the Soxhlet extraction method. The formulated soybean mud properties were compared with diesel oil mud properties. The compared properties were rheological properties, yield point and gel strength, and mud density and filtration loss properties, fluid loss and filter cake. The results obtained show that the soybean oil mud exhibited Bingham plastic rheological model with applicable (low) yield point and gel strength when compared with the diesel oil mud. The mud density measurement showed that soybean OBM was slightly higher than diesel OBM with mud density values of 8.10 lb/gal and 7.98 lb/gal, respectively, at barite content of 10 g. Additionally, the filtration loss test results showed that soybean mud fluid loss volumes, water and oil, were 13 mL and 10 mL, respectively, compared to diesel oil mud volume of 15 mL and 12 mL. Furthermore, the filtration loss test indicated that the soybean oil mud with filter cake thickness of 2 mm had a cake characteristic of thin and soft while the diesel oil mud resulted in filter cake thickness of 2.5 mm with cake characteristic of firm and rubbery. In comparison with previous published works in the literature, the soybean oil mud exhibits superior rheological and filtration property over other vegetable oil-based muds. Therefore, the formulated soybean oil mud exhibited good drilling mud properties that would compare favourably with those of diesel oil muds. Its filter cake characteristic of thin and soft is desirable and significant to avert stuck pipe during drilling operations, meaning that an oil-based drilling mud could be formulated from soybean oil.

INTRODUCTION

The need to drill a usable hole with minimal environmental impact and with a low cost imprint has been a dream long held by drillers and industry operators alike. One major component of the oil well drilling operation which is often referred to as "the blood of the drilling process" is the drilling fluid. It is the "architect" which can make the drilling operation either materialize or unrealizable. This is because the fluid plays a number of invaluable roles. These roles include but are not limited to transporting the drill cuttings from the bottom of the hole to the surface, cooling and lubricating the drilling bit as well as the drill string to minimize its wear, sealing off permeable formations by forming an impermeable, relatively thin mud cake at the borehole wall of the permeable formations, creating an overbalanced drilling condition to control the formation pressure, hold drill cuttings in suspension when circulation is interrupted, creating a buoyancy force to partly support the weight of the drill string and casing string, reduction of formation damage of various horizons penetrated, transmission of hydraulic horsepower to the bit and allowing maximum penetration rates, carrying downhole information from the drilled well in form of signals to the surface for interpretation, and so forth. Therefore, the success of any rotary drilling operation is hinged on the performance of the drilling fluid used for the drilling operation. In turn, the performance of these drilling fluids is dependent on the rheological properties of the mud used. These rheological properties include plastic viscosity, yield point, and gel strength, among others. Drilling fluid costs are estimated to gulp about 20% of the total drilling cost of a well [1]. As explorationists make discoveries in unconventional terrains such as deep water offshore fields, ultra high temperature, high pressure fields, arctic regions, and other hostile environments, the cost of drilling for oil and gas reserves becomes more expensive. From a drilling fluids perspective, the demands of intrinsic to deep-water drilling and completions are especially acute as operators must reconcile performance and economic objectives with unique technical and environmental obstacles [2]. These drilling environments require fluids that excel in performance. Therefore, measuring fluid performance requires the evaluation of all key drilling parameters and their associated cost [3]. Over time, the oil well drilling industry has basically made use of two types of drilling fluids, namely, water-based muds (WBMs) and oil-based drilling muds (OBMs). On the one hand, due to the lower costs and ease of formulation, water-based mud is most commonly used as drilling fluid. On the other hand, the oil-based muds despite being more costly when compared with their water based counterparts are used because of their good rheological characteristics at temperatures as high as 500°F exhibiting better stability behaviour, their effectiveness against all types of corrosion, and superior lubricating characteristics. Oil-based muds have the additional advantage of being able to drill through formations containing water swellable clays. To formulate these oil-based muds, diesel oil is used as the base fluid primarily because of its viscosity characteristics, low flammability, and low solvency for rubber [4]. All such petroleum-based oils used for drilling mud contain relatively large amounts of aromatics and at least a substantial concentration of n-olefins both of which may be harmful or toxic to animal and plant life [5]. As such, the drilling industry has over time developed variant forms of oil-based muds which are technically called synthetic-based muds (SBMs). An SBM was used for the first time to drill a well in the Norwegian Sector of the North Sea in 1990. The first well drilled with an SBM in the UK Sector was in 1991 and in the Gulf of Mexico in 1992 [6, 7]. These SBMs combine the desirable operating qualities of the oil-based mud and lower the toxicity and environmental impact qualities of the water-based mud. In this paper, two oil-based muds are formulated and compared; one is formulated with diesel oil and the other from soybean oil. The basis for the comparison stems from the rheological properties and the filtration characteristics of the drilling muds.

MATERIALS AND METHODS

Sample Preparation

The soybean seeds were collected from a local market in Uyo, Akwa Ibom State, a state in the southern part of Nigeria. The outer skin was first removed and then it was ground using a manual grinding machine, before finally extracting oil from them. The method used in extracting oil from the seeds was the solvent extraction method. This method involves extracting oil from oil-bearing materials by treating it with a low boiling point solvent. This method was preferred to all other extraction methods (such as expellers and hydraulic presses) because it recovers almost all the oil and leaves behind only 0.5% to 0.7% residual oil in the raw material [8, 9]. Ten (10) kg of ground soybean seeds were measured out, tied in a filter paper, and loaded into the extraction chamber of the Soxhlet extractor. About 500 mL of n-Hexane was poured into the round bottom flask (i.e., the boiling flask) contained in the heating mantle of the system, and the heating mantle was turned on and allowed to heat at 60°C. The solvent was heated to reflux and the solvent vapour moved up the distillation arm and entered into the condenser which condenses the vapour. At this point, the condensed vapour dripped back down into the extraction chamber housing the samples. Then at a certain level, the siphon emptied the liquid in the extraction chamber into the boiling flask. This cycle continued until the liquid in the flask changed colour (yellow in this case) to a considerable extent. At this point, the fluid mixture (n-hexane and soya oil) was collected in a glass beaker and was separated using a steam bath. The steaming was done at 60°C; the hexane evaporated while the extracted oil remained in the beaker and was collected and stored. This process was repeated until enough quantity of the soybean oil was obtained for the mud formulation.

Physicochemical Properties of the Oil-Based Fluids

The fluids used as base-fluid in the formulation of the OBM were analyzed to determine its physiochemical properties. By knowing these parameters, an early description of mud composition and behaviour is estimated. Therefore, the parameters that were tested include the following.

- Specific gravity: it shows the weight of base-fluid. This will indicate the density of developed mud.
- Pour point: it shows the lowest temperature at which the base-fluid will be able to flow.
- Flash point: it shows the temperature at which the fluid begins to burn.
- Kinematic viscosity: it shows the resistance of base fluid to flow under the influence gravity force.
- Cloud point: this is the temperature at which dissolved solids are no longer completely soluble, precipitating as a second phase giving the fluid a cloudy appearance.

Interestingly, some research work has been done by researchers to determine the aforementioned base fluid properties. Table 1 presents the physiochemical properties of diesel oil and soybean oils.

Table 1: Physicochemical properties of diesel and soybean oil

Property	Diesel oil # 2	Soybean oil
Physical form	Liquid	Liquid
Melting point	−30°C to −18°C[a]	−16°C[b]
Boiling point	282°C–338°C[a]	257°C[d]
Density	830 kg/m³ at 20°C[c]	920 kg/m³ at 20°C[c]
Water solubility	Immiscible	Immiscible
Flash point	70°C[c]	330°C[c]
Cloud point	−9°C[c]	−4°C[c]
Kinematic Viscosity	6 Cst at 20°C[c]	61 Cst at 20°C[c]
Pour point	−18°C[c]	−20°C[c]

Source:

[a]http://www.inchem.org.

[b]http://www.engineeringtoolbox.com.

[c]http://www.proceedings.scielo.br/img/eventos/agrener/n4v1/064t01.gif.

[d]http://van.physics.illinois.edu/qa/listing.php?id=1428.

Chemical Stability of Soybean Oil and Diesel Oil

Aside from knowing the physiochemical properties of the soybean oil, the knowledge of its chemical stability at wellbore condition (high temperature) becomes paramount for its use as base fluid in OBMs formulation. According to Wikipedia (online), a chemical substance is said to be "stable" if it is not particularly reactive in the environment or during normal use and retains its useful properties on the timescale of its expected usefulness. In particular, the usefulness is retained in the presence of air, moisture, or heat and under the expected conditions of application. In this direction, the biodegradability provides an indication of the persistence of any particular substance in the environment and is the yardstick for assessing the eco friendliness of substances [13]. So many studies by researchers have shown unanimity as to the biodegradable nature of soybean oil and other vegetable oils as compared to diesel oil. For instance, Howell [14] and Erhan and Perez [15] all agree that soybean oil and other vegetable oils are environmentally friendly, renewable, nontoxic, and biodegradable. To further buttress the benefits of soybean oil as a viable alternative to the use of diesel oil in OBM formulation from the chemical stability standpoint, Table 2 presents some of the properties for this assertion. The table presents the base oil requirements for use as oil base in OBMs formulation. In comparison, diesel oil exhibits high volatilities with low flash points which may lead to safety hazards compared to soybean oil with high flash point and low volatility. Thus, diesel oil when used as OBMs would jeopardize the health and safety of its users and the environment as well.

Table 2: Base properties of oil for use as OBMs

Property	Aniline point (°C)	Pour point (°C)	Flash point (°C)	Fire point (°C)	Kinematic viscosity at 40°C, Cst	Aromatic content (%)
Base oil required properties	>65	<Ambient temperature	>66	>80	2.3–3.5	4–8
Diesel oil	145	−18	66	108	2.7–3.4	25% v/v
Soybean oil	60	−20	330	342	31.8	N/A

Source: Yassin et al., 1991 [10].

In addition, the pour and fire points of diesel oil compared to that of soybean oil imply that the former would be chemically unstable at extreme temperatures (i.e., low and high), as diesel oil flow rate at −18°C becomes difficult. Also, in high temperature-high pressure (HTHP) well, diesel oil with fire point of 108°C will ignite, thus making its use as base fluid in OBMs detrimental. However, the only draw-back for the soybean oil is the low aniline point which could be detrimental to rubber elements on the drilling rig.

Mud Formulation

Formulations of oil-based mud with some additives were performed in this study. The following additives were used for the developed oil-based muds: primary and secondary emulsifier (GLO PEMUL 1000 and GLO SEMUL 1000, resp.), filtration control additive (hydroxy ethyl cellulose), viscosifier (bentonite), weighting material (barite), and caustic soda (NaOH) along with (pH enhancer). For comparison, diesel oil and soybean oil were used to formulate oil-based drilling fluid using same formulation specifications based on API (American Petroleum Institute) standard of 25 g bentonite to 350 mL base fluid for nontreated bentonite. Thus, the mud samples were formulated based on the concept of maintaining the same component proportions in each fluid. The oil-water ratio used in formulating the mud was 70 to 30. The mixing order used during the formulation of the two muds is presented in Table 3. The formulated drilling muds were allowed to age for 24 hours before testing for their various mud properties. Thereafter, barite content of 10 g was added to the formulated muds to observe their mud density behaviour to barite content.

Table 3: Composition of soybean oil and diesel oil mud

Mud component	Diesel oil # 2	Soybean oil	Mixing duration (mins)	Mixing order
Oil (mL)	245	245	—	1
Primary emulsifier (mL)	6	6	5	2
Secondary emulsifier (mL)	4	4	5	3
Filter loss agent HEC (g)	0.35	0.35	5	4
Water (mL)	105	105	15	5
Bentonite (g)	25	25	5	6
NaOH (g)	0.25	0.25	5	7
Barite (g)	10	10	10	8

Mud Properties Determination

Rheological Properties Measurement

The Fann V-G viscometer was used to determine the rheological properties of the mud samples. The equipment was switched on and allowed to stabilize, after which the viscosity of distilled water was tested to check the integrity of the equipment. The mud sample was poured into the cup of the viscometer and was placed on the viscometer stand. The stand was adjusted and held in position as the rotor sleeve was immersed in the mud exactly to the fill line. The speed selector knob was selected to rotate at 600 rpm (revolutions per minute), and the power was switched on. When a steady dial reading was reached, this was recorded as the 600 RPM dial reading. The above process was repeated for 300, 200, 100, 6, and 3 rpms. The rheogram (shear stress-shear rate profile) for the readings is presented in Figure 1. Consequently, the apparent viscosity (AV), plastic viscosity (PV), and yield point (YP) of the mud samples (presented in Table5) were calculated based on two data point approach from the viscometer (Fann V-G) dial readings presented in Table 4 using

$$PV = \theta_{600} - \theta_{300},$$ (1)

$$YP = \theta_{300} - PV,$$ (2)

$$AV = \frac{1}{2} * \theta_{600},$$ (3)

where AV is apparent viscosity, PV is plastic viscosity, θ_{600} is dial reading at 600 revolutions per minute, θ_{300} is dial reading at 300 revolutions per minute, and YP is yield point.

Table 4: Viscometer readings of the formulated mud samples

Viscometer speed (rpm)	Diesel oil		Soybean oil	
	Dial reading	Shear stress (lb/100 ft²)	Dial reading	Shear stress (lb/100 ft²)
600	51	54.06	38	40.28
300	35	37.1	24	25.44
200	27	28.62	16	16.96
100	18	19.08	9	9.54
6	9	9.54	3	3.18
3	7	7.42	2.3	2.438

Table 5: Mud rheological parameters value

Mud properties	Diesel oil	Soybean oil
Gel strength 10 sec, lb/100 ft²	7	5
Gel strength 10 min, lb/100 ft²	9	6
Plastic viscosity, cp	16	14
Yield point, lb/100 ft²	20	11
Apparent viscosity, cp	25.5	19

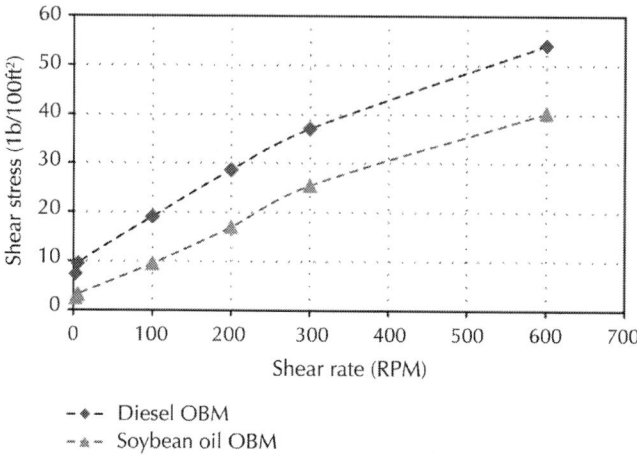

Figure 1: Shear stress-shear rate profile.

The gel strength of the mud samples were also determined using the Fann V-G viscometer. The speed selector knob was rotated to stir the mud sample for ten seconds, and then it was rotated at 3 rpm and the power was immediately shut off. As soon as the sleeve stopped rotating, the power was turned on after 10 seconds and 10 minutes, respectively, at 3 rpm. The maximum dial was recorded for each case as the gel strength of the mud samples at 10 sec and 10 min. Thus, the results are presented in Table 5 and Figure 2.

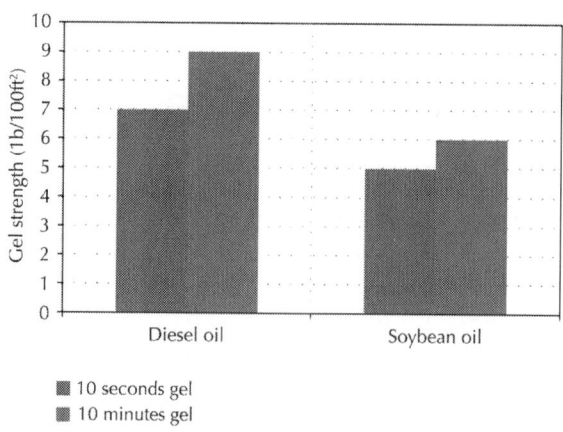

Figure 2: Gel strength of diesel oil and soybean oil OBMs.

Mud Density Measurement

A mud balance was used to determine the mud density. The mud balance was calibrated by using it to test for the density of water before testing for the densities of the mud samples. The mud balance cup was filled to the top with the mud samples. The lid was placed on the cup and was turned to ensure that it was firmly put in place. Excess mud spilled through the vent was wiped off from the lid. The balance was placed on a knife edge and the rider was moved along the graduated arm until the cup and the arm were balanced as indicated by the bubble. The mud weight in pounds per gallon (lb/gal) was read at the edge of the rider towards the mud cup as indicated by the arrow on the rider and recorded. The muds' density without barite content and with barite content of 10 g were measured and recorded as presented in Table 6.

Table 6: Mud density values with barite content

Barite (g)	Mud density (lb/gal)	
	Diesel oil	Soybean oil
0	7.50	7.82
10	7.98	8.10

Mud Filtration Properties Measurement

The filtration test was performed using standard cell at API condition of 100 psi differential pressure at room temperature. The filter press used for the tests consisted of six independent filter cells mounted on a common frame (as shown in Figure 8(b)) after which the mud to be tested was introduced into the cup assembly. With the air pressure valve closed, the mud cup assembly was clamped to the frame while holding the filtrate outlet end finger tight. A graduated cylinder was placed underneath to collect filtrate, after which the pressure valve was opened for gas to flow in from an air compressor pump and the timing was started at the same time. After 30 minutes the pressure valve was shut and the filtrate collected in the graduated cylinder (in mL) and the filter cake (in mm) were measured and recorded. The results of the test are presented in Table 7 and Figures 3 and 4.

Table 7: Mud filtration loss results

Filtrate properties	Diesel OBM	Soybean OBM
Total fluid volume (mL)	27	23
Oil volume (mL)	12	10
Water volume (mL)	15	13
Cake thickness (mm)	2.5	2.0

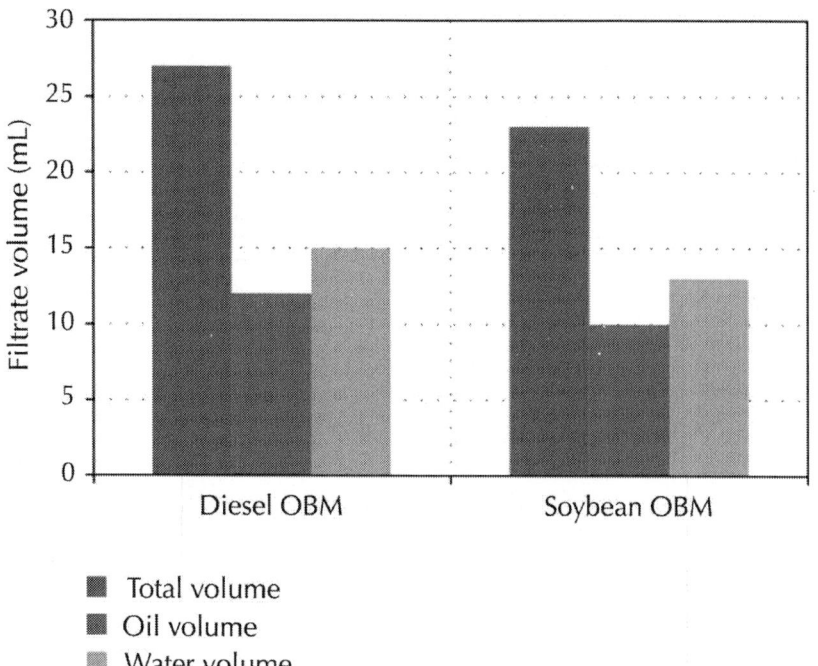

Figure 3: Filtration volume plot for diesel and soybean OBMs

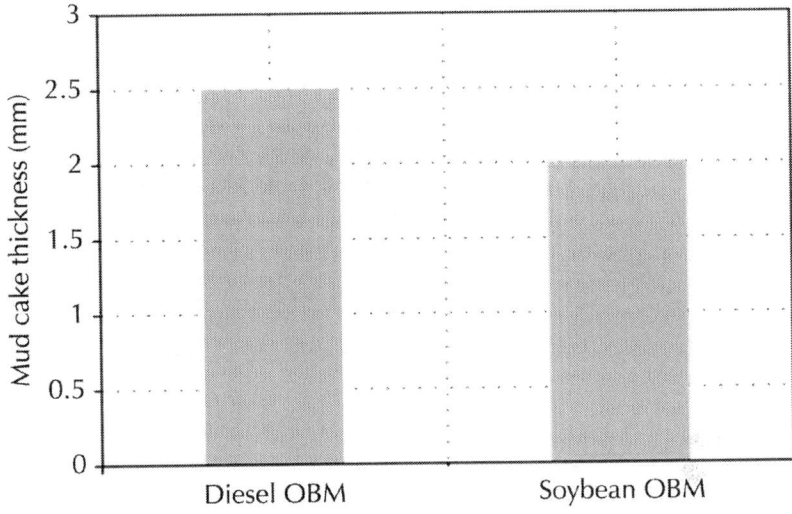

Figure 4: Comparison of mud cake thickness.

Additionally, calculation of filtrate loss at variable time intervals relative to known filtrate loss and time interval can be predicted using the expanded equation

$$f_1 = f \times \left(\frac{\sqrt{T_1}}{\sqrt{T}} \right),$$

(4)

where f is known filtrate at a time interval of T and f_1 is unknown filtrate at a time interval of T_1.

Therefore, the calculated values for filtration volume at times of 1, 5, 7.5, 10, and 15 minutes are presented in Table 8 and depicted in Figures 5 and 6. In addition, the spurt loss (initial loss) volume from the formulated mud was estimated using the expanded equation

$$V_{sp} = V_{1.0} - \left[\frac{V_{7.5} - V_{1.0}}{\sqrt{7.5} - \sqrt{1}} \right] \sqrt{1},$$

(5)

where V_{sp} is spurt loss volume, $V_{1.0}$ is filtrate volume obtained at time of one (1) minute, and $V_{7.5}$ is filtrate volume obtained at time interval of 7.5 minutes.

Table 8: Calculated mud filtration

Mud type	Time (minutes)					
	1	5	7.5	10	15	30
	Filtrate volume (mL)					
Diesel OBM	4.93	11.02	13.5	15.58	19.09	27
Soybean oil OBM	4.20	9.39	11.5	13.27	16.26	23

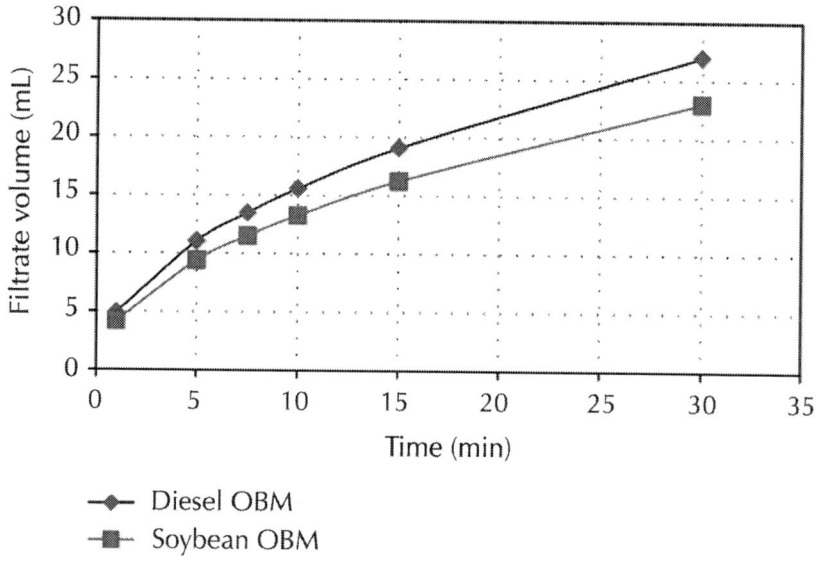

Figure 5: Comparison of filtrate loss with time.

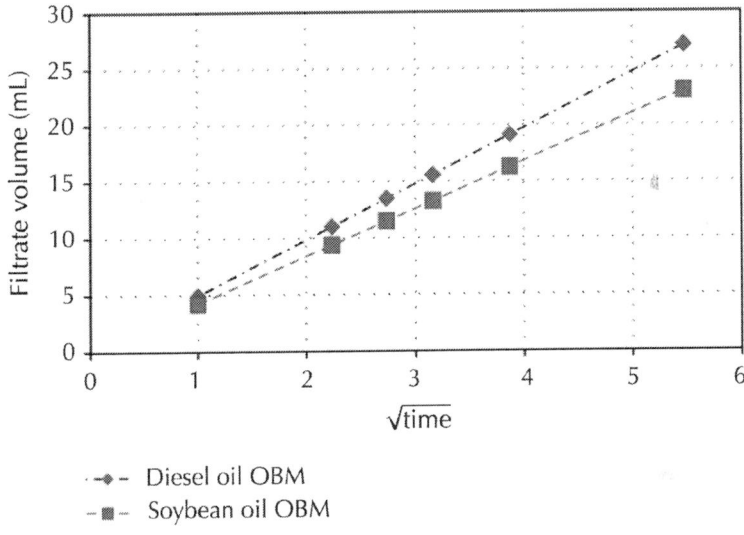

Figure 6: Comparison of filtrate volume versus square root of time.

DISCUSSION OF RESULTS

Rheological Properties

The fundamental reason for choosing to study the rheological properties, plastic viscosity, yield point, and gel strength, as well as the filtration properties, fluid loss and filter cake of both oils as the basis for comparison, is the relevance these properties offer to the overall drilling mud performance. The yield point (YP) is used to evaluate the ability of a mud to lift cuttings out of the annulus. A high YP implies a non-Newtonian fluid; one that carries cuttings better than a fluid of similar density but lower YP. Additionally, frictional pressure loss is directly related to the YP. It is important to state here that excessively high YP leads to high pressure losses while the drilling mud is being circulated. The viscometer speeds and dial readings were converted to shear rate and shear stress, respectively. Figure 1 presents the shear stress-shear rate profile (rheogram) of the formulated mud from the values obtained. This figure depicts the rheological model of the two mud samples. As shown in the figure, the rheology of both mud samples is similar (i.e., increasing from left to right with intercept on the vertical axis). This indicates that the mud samples are similar in their rheological behavior. In this connection, it can be seen that the two formulated OBMs exhibit a rheological model which is almost similar to the Bingham plastic model. A Bingham plastic fluid will not flow until the shear stress (τ) exceeds a certain minimum value known as the yield point, YP [4]. After the yield point (YP) has been exceeded, the changes in shear stress are proportional to changes in shear rate in which the constant of proportionality is known as the plastic viscosity (PV). From Figure 1, diesel OBM has a higher plastic viscosity compared to soybean OBM. This indicates that diesel OBM has a higher viscosity, which would offer a greater resistance to fluid flow that will result in increased circulating pressures that can cause loss of circulation and increased pumping costs. Thus, soybean OBM with low viscosity is a good prospect for drilling in the sense that its low viscosity will offer less resistance to fluid flow and therefore would lead to a turbulent flow at low pump pressure which would result in good hole cleaning.

The gel strength is another important drilling fluid property, as it demonstrates the ability of the drilling mud to suspend drill solid (drilled cuttings) and weighting material when mud circulation is ceased. Figure 2 depicts the gel strength of the formulated diesel and soybean OBMs. The figure and Table 4 indicate that soybean OBM has low gel strength values at both 10 seconds and 10 minutes compared to diesel OBM. Thus, the gel strength values of the soybean oil-based mud show that the mud exhibits a flat gel structure, meaning that the mud will remain pumpable with time if left static in the hole. Conversely, the gel strength values of the diesel oil-based mud at 10-minute gel value were much higher than the 10-second gel value, indicating that the diesel mud exhibits a progressive gel structure. This is an indication that the gelation of the diesel mud is rapidly gaining strength with time, which generally is an undesirable feature of a drilling mud. Therefore, the weak gel/fragile property of the soybean OBM is desirable during drilling operation as the gel can be broken easily with lower pump pressure to make circulation. On the other hand, the high value of diesel OBM gel strength would lead to high circulation breakdown pressure and increased pumping costs as high pump power is required to overcome this gelling potential of diesel OBM.

Mud Density

Mud density of drilling fluid system is mainly necessary for the control of formation pressures. Additionally, an increase in mud density increases the capacity of the mud to carry drilled cuttings as the suspending fluid has an associated buoyancy effect on the cuttings. As presented in Table 6, soybean mud without barite content has higher mud density (7.82 lb/gal) than diesel mud (7.50 lb/gal), an indication owing to the soybean oil density (920 kg/m³ about 7.66 lb/gal) as presented in Table 1. Table 6 further depicts the mud's density variation at 10 g barite content. From the table, the formulated soybean OBM has higher density value than diesel OBM. The results indicate an increase in soybean mud density to 8.10 lb/gal and diesel mud density to 7.98 lb/gal at 10 g barite content. These results indicate an increase of 3.58% and 6.40% for soybean and diesel OBMs' density, respectively. This assertion indicates that soybean OBMs density can be increased with barite to suit any drilling operation especially when it is required that a high mud weight is necessary. A move that is very significant in the event of hole drilling problems and when gas cut mud is encountered.

Filtration Loss

Filtration rate is often the most important property of a drilling fluid, particularly when drilling permeable formations where the hydrostatic pressure exceeds the formation pressure. Proper control of filtration can prevent or minimize wall sticking and drag and in some areas improve borehole stability. Figure 3 and Table7 show filtrate volume from the formulated OBMs collected after 30 minutes. The results depict that the water volume collected from diesel mud 15 mL was higher than the soybean mud volume of 13 mL. This could be attributed to the fact that the water may not have been completely emulsified in the diesel mud, thereby forming an unstable emulsion. However, the volumes of oil collected from both samples were 12 mL and 10 mL for diesel and soybean mud, respectively. Additionally, Figure 5 shows the calculated filtrate loss volumes for time interval of 1.0, 5.0, 7.5, 10.0, 15.0, and 30.0 minutes. The figure indicates that the filter loss values increased with time, as the diesel OBM shows high filter loss capacities compared to the soybean OBM. Figure 6 presents the filtrate volumes against the square root of the time in minutes. This move was to ascertain the spurt loss of the formulated OBMs. Thus, the spurt loss values obtained for the two mud samples were, however, the same with spurt loss volume of about 0.01 mL.

Mud Cake Thickness

As earlier alluded to the diesel OBM has a higher filtration rate than the formulated soybean OBMs. Generally, high filtrate volumes are associated with thick filter cake because the cake is formed by deposition of clay particles on the walls of the hole during filtrate loss to the formation. So the higher the filter volume, the thicker the filter cake and the less efficient the drilling mud. In view of its high filtration rate, diesel OBM has a thicker mud cake (2.5 mm) than soybean OBM (2.0 mm) as presented in Figure 4. The effect of this is that a thick mud cake reduces the effective diameter of the drilled wellbore thereby increasing the area of contact between the drill pipe and the cake leading to increased risk of stuck pipe incidents. Based on this result, therefore, the formulated soybean OBM has good filtration properties that will be effective for drilling purposes as it would prevent stuck pipe incidents. In addition, diesel OBM filter cake has a firm and rubbery physical characteristic while the soybean OBM has a soft physical characteristic as indicated in Figure 8(d). On this note, the thin but soft mud cake obtained in the case of the soybean oil OBM is desirable during drilling operations.

Comparison of Soybean Oil Mud with Previous Works on Other Oil Base Fluids

In the literature, limited or no direct comparison of the rheological properties of soybean oil with other vegetable oils as base fluid in oil-based drilling mud formulation is available. However, the literature presents the comparison of different vegetable oil as alternative to diesel oil in oil-based mud formulation. Flowing from this, Adesina et al. [8, 9] presented algae, moringa, canola, and jatropha oils as alternative to diesel oil in oil-based mud formulation. Their study as presented in Table 9 revealed the following mud properties for the various oils they studied.

Table 9: Previous works (Adesina et al. [8, 9]) mud properties

Mud properties	Diesel oil	Canola oil	Jatropha oil	Algae oil	Moringa oil
Density, lb/gal	8.26	8.47	8.32	7.81	8.30
pH	8.00	9.50	9.00	9.00	9.00
Rheological properties					
Plastic viscosity, cP	13.0	12.0	8.0	8.0	11.0
Apparent viscosity, cP	92.5	64.0	77.0	61.0	84.5
Yield point, lb/100 ft^2	155	112	112	106	147
Gel strength, lb/100 ft^2	50/51	60/72	54/55	52/43	52/53
Filtration property					
Total fluid volume, mL	6.90	6.00	6.30	6.20	7.20
Oil volume, mL	2.30	1.00	1.10	1.10	2.50
Water volume, mL	4.60	5.00	5.20	5.10	4.70
Cake thickness, mm	1.00	0.90	0.89	0.90	0.90

Source: Adesina et al., 2012 [8, 9].

Furthermore, Paul et al. [11] evaluated the viability of jatropha and groundnut oils as alternative to diesel oil in oil-based drilling mud formulation. The obtained mud properties from their study are shown in Table 10.

Table 10: Previous works (Paul et al. [11]) mud properties

Mud properties	Diesel oil	Jatropha oil	Groundnut oil
Density, lb/gal	7.5	8.5	7.9
Plastic viscosity, cP	8.0	15	35
Apparent viscosity, cP	25	55	42.5
Yield point, lb/100 ft^2	32	45	15
Gel strength, lb/100 ft^2	16/15	8/10	4/7

Source: Paul et al., 2014 [11].

Additionally, Akintola et al. [12] examined the use of locally sourced oil: groundnut oil, melon oil, vegetable oil, soya bean oil, and palm oil as substitute for diesel oil in formulating oil base drilling fluids. Their work focused on evaluating the filtration property of the aforementioned oils and comparing the result with formulated diesel oil mud's filtration property. The pH and filtration test result as obtained from their study is presented in Table 11.

Table 11: Previous works (Akintola et al. [12]) mud properties

Mud properties	Diesel oil	Groundnut oil	Melon oil	Palm oil	Soya oil	Vegetable oil
pH	9.0	8.0	8.0	8.0	8.0	8.0
Filtrate volume, mL	14.5	12.5	15.0	11.0	13.5	15.0

Source: Akintola et al., 2014 [12].

Actually, the yardstick of comparison of the previously published works on various vegetable oil with the soybean oil as base fluid in mud formulation becomes worrisome, as various authors used different mud formulation standards: API and OCMA standard. Adesina et al. [8, 9] did not state their drilling mud formulation standard whereas Paul et al. [11] used API standard of 20 g bentonite to 350 mL liquid phase: oil-water ratio of 70 : 30, with addition of 106 g of weighting material (barite) to the formulated mud. Also, Akintola et al. [12] used 350 mL continuous phase of 70 : 30 oil-water ratio with unstated grams (g) of bentonite. Hence, aside from the addition of barite, Paul et al. [11] and this study used almost the same drilling mud formulation standard. Thence, the comparison of the various published work results, Tables 9through 11, with this study results, Tables 5 and 7, indicates that favourable rheological and filtration properties would be achieved with soybean oil over jatropha, canola, and moringa oil, among others, as base fluid in oil-based drilling mud formulation. Owing to the fact that, the previously published vegetable oils resulted in high rheological properties, plastic viscosity, yield point, and gel strength which will require high pump power to initiate mud circulation during drilling operation.

Cost Comparison of Soybean Oil and Diesel Oil

For every business venture, an economic analysis or a cost benefit analysis of the project to be undertaken is required. In the same vein, in carrying out a comparative study of the use of soybean oil as substitute for diesel oil in drilling mud formulation, it is imperative that the cost of these oils is factored into consideration. Thus, Table 12 and Figure 7 show the price of both oils in the last six years.

Table 12: Price of soybean oil and diesel oil in the last six years

Year	Soybean oil ($/bbl)	Diesel oil ($/bbl)
2009	119.58	81.48
2010	135.39	92.22
2011	177.98	126.63
2012	168.61	130.76
2013	147.98	126.66
2014	123.68	124.46

Source: Index Mundi (online) Commodity prices. http://www.indexmundi.com/commodities/.

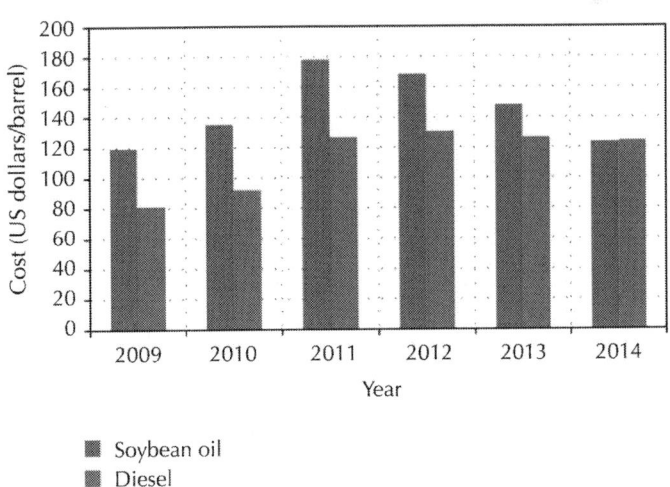

Figure 7: Trend of price of soybean oil and diesel oil in the last six years.

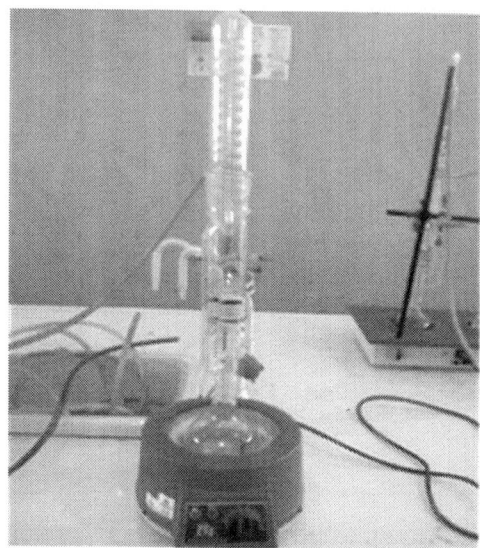

(a) Soxhlet extractor

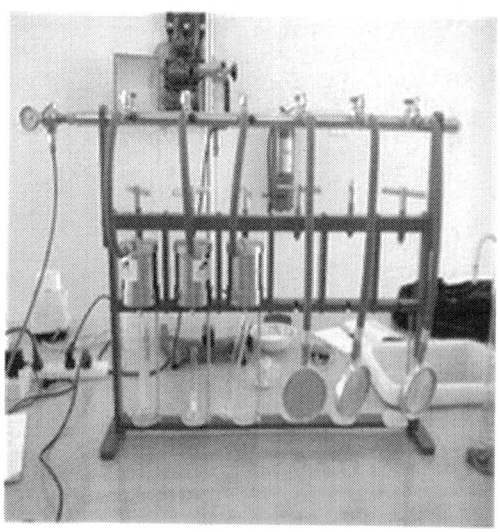

(b) API Filter press

(c) Ground soybean in API filter paper

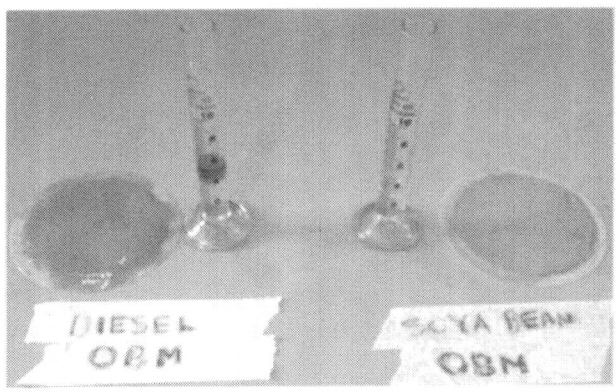

(d) Characteristics of the filter cake of the formulated muds

Figure 8: Equipment and materials used during test.

From the Table 12 and Figure 7, it is observed that the cost per barrel of soybean oil in the previous five years (i.e., 2009 through 2013) remained higher than that of the number 2 diesel oil. Interestingly, in 2014, the cost per barrel of soybean oil became slightly lower than diesel oil as its production increased. Although the high cost of soybean oil would definitely increase the cost of formulating OBMs with it, this high cost, however, would eventually be offset by the superior rheological properties of the soybean oil. Additionally, soybean oil is environmentally friendly (biodegradability) in terms of cuttings disposal and reduced liabilities in the event of spillage.

CONCLUSIONS

It is a well-known fact that the performance of drilling fluid during drilling operation is influenced by its properties such as mud viscosity, density, pH, and filtration loss, among others. In this study, soybean oil was used as a base fluid in oil-based mud (OBM) formulation. The formulated soybean OBM properties were compared with diesel OBM. While the formulated soybean OBM has very good potentials as oil-based drilling mud when compared with diesel OBM, based on the results obtained from the study, the following conclusions can be drawn.

- The formulated soybean OBM has a Bingham plastic rheological model with low yield point and gel strength, mud property desirable for turbulent flow at low pump pressure for effective hole cleaning.
- The soybean OBM has relatively high density and can be increased with densifiers to desirable values during equivalent circulating density (ECD) predictions in order to obtain a successful drilling operation.
- The filtration loss property of the formulated soybean OBM compared favorably with diesel OBM with filter cake characteristic of thin and soft desired during drilling operations.

REFERENCES

1.   Shell Petroleum Development Company, Shell Intensive Training Manual, Shell Petroleum Development Company, 2000.

2.   J. E. Friedheim, "Area-specific analysis reflects impact of new generation fluid systems on deepwater exploration," in Proceedings of the IADC/SPE Asia Pacific Drilling Technology Conference, IADC/SPE 47842, Society of Petroleum Engineers, Jakarta, Indonesia, September 1998. ·

3. B. Hughes, Drilling Fluids Reference Manual, 2006.

4. A. T. Bourgoyne, M. E. Chenevert, K. K. Millheim, and F. S. Young, Applied Drilling Engineering, vol. 2 of SPE Textbook Series, SPE, Richardson, Tex, USA, 2003.

5. M. M. Dardira, S. Ibrahimea, M. Solimanb, S. D. Desoukya, and A. A. Hafiza, "Preparation and evaluation of some esteramides as synthetic based drilling fluids," Egypt Journal of Petroleum, vol. 23, no. 1, pp. 35–43, 2014.

6. J. E. Friedheim and H. L. Conn, "Second generation synthetic fluids in the north sea: are they better?" in Proceedings of the SPE/IADC Drilling Conference, IADC/SPE 35061, pp. 215–228, Society of Petroleum Engineers, New Orleans, La, USA, March 1996.

7. R. G. Fechhelm, B. J. Gallaway, and J. M. Farmer, "Deepwater sampling at a synthetic drilling mud discharge site on the outer continental shelf, Northern Gulf of Mexico," in Proceedings of the SPE/EPA Exploration and Production Environmental Conference, SPE 52744, pp. 509–513, Society of Petroleum Engineers, Austin, Tex, USA, February-March 1999. ·

8. F. Adesina, A. Anthony, A. Gbadegesin, O. Eseoghene, and A. Oyakhire, "Environmental impact evaluation of a safe drilling mud," in Proceedings of the SPE Middle East Health, Safety, Security and Environment Conference, SPE 152865, pp. 2–4, Abu Dhabi, United Arab Emirates, April 2012.

9. F. Adesina, F. Olugbenga, A. Churchill, A. Abiodun, and A. Anthony, "Novel formulation of environmentally friendly oil based drilling mud," in New Technologies in the Oil and Gas Industry, chapter 3, InTech, 2012. ·

10. A. M. Yassin, A. Kamis, and M. O. Abdullah, "Palm oil diesel as a base fluid in formulating oil based drilling fluid," SPE Paper 23001, Society of Petroleum Engineers, 1991.

11. A. L. A. Paul, V. E. Efeovbokhan, A. A. Ayoola, and O. A. Akpanobong, "Investigating alternatives to diesel in oil based drilling mud formulations used in the oil industry," Journal of Environment and Earth Science, vol. 4, no. 14, pp. 70–77, 2014.

12. S. Akintola, A. B. Oriji, and M. Momodu, "Analysis of filtration properties of locally sourced base oil for the formulation of oil

based drilling fluids," Scientia Africana, vol. 13, no. 1, pp. 171–177, 2014.

13. E. O. Aluyor, K. O. Obahiagbon, and M. Ori-Jesu, "Biodegradation of vegetable oils: a review,"Scientific Research and Essays, vol. 4, no. 6, pp. 543–548, 2009. ·

14. S. Howell, Promising Industrial Applications for Soybean Oil in the US, American Soybean Association, National Biodiesel Board, 2007.

15. S. Z. Erhan and J. M. Perez, Biobased Industrial Fluids and Lubricants, The American Oil Chemists' Society, 2002.

Chapter 5

Correlation and Prediction of the Solubility of Solid Solutes in Chemically Diverse Supercritical Fluids Based on the Expanded Liquid Theory

Loubna Nasri[1], Salima Bensaad[2], and Zouhir Bensetiti[3]

[1]Department of Pharmaceutical Engineering, University Constantine 3, Constantine, Algeria

[2]Department of Chemistry, University Constantine 1, Constantine, Algeria

[3]Unisignal Inc., Brossard, Canada

ABSTRACT

For the proper design of any extraction procedure based on supercritical solvents, it is essential to have a sound knowledge of the solubility data of different compounds and the accurate way to represent it. The solute's solubility in a supercritical solvent is dependent on the solute, the solvent, and the operating conditions (temperature and pressure). Developing a comprehensive experimental data set is an onerous task and time consuming and, thus, the incentive to develop predictive tools is substantial. In this paper, a technique is presented and tested to correlate and predict solute's solubility in different supercritical fluids with a methodology based on the expanded liquid theory, in which the solid-fluid equilibrium is modeled using the local composition model of UNIQUAC in which the interaction parameters are related to the solvent reduced density with empiric equations. The most advantages of this model include: it does not require the knowledge of critical properties and sublimation pressure of solid solutes and does take into account the binary interaction between solid solute and solvent. The evaluation of the proposed model capabilities is done by testing it on a large data base consisting of experimental solubility data taken from literature of 33 binary systems solid-SC fluid. The results obtained for both correlation and prediction show good agreement with the experimental data used. For the comparison we have considered some literature models that account for effect of the system conditions (temperature and pressure) in addition to the sublimation pressure of the solute through their introduction of the enhancement factor and a model based on a modified Peng-Robinson equation of state.

INTRODUCTION

During the past few years, widespread attention has been focused on supercritical fluids due to their potential application in extraction processes in foods processing, pharmaceuticals, flavors, chemicals and petroleum industries. The main advantages of supercritical fluid extraction over conventional extraction methods include increased speed, easy solvent separation and better recovery, and reduction in both solvent usage and solvent waste generation. The solubility of a solute in a supercritical fluid is perhaps the most important thermophysical property that must be determined and modeled for an efficient design of any extraction procedure based on supercritical solvents. The determination of solubilities of a wide variety of solids and liquids of low volatility in supercritical fluids has received considerable attention in recent years. However, despite the vital importance of the solubility data of different compounds from chemical, biochemical, pharmaceutical and industrial points of view, there is still a lack of fundamental solubility and mass-transfer data available in the literature to facilitate the development of commercial-scale processes. Since the experimental determination of the solubilities of various solutes in supercritical fluids at each operating condition is tedious, time-consuming and not reported in literatures, there is a considerable interest in mathematical models that can accurately predict the solubilities of solid solutes in supercritical fluids [1]. Therefore it is essential to have a model that not only can accurately correlate but also predict phase equilibrium properties. Some of the models that have been used for correlating solubility data can be classified in two classes, equations of state based models (EOS) [2] and empirical models [3,4]. EOS based models require the prior knowledge of a certain number of parameters such as the critical properties (temperature and pressure), acentric factor and the sublimation pressure of the solid solute. These parameters are not available and specifically for many high molecular weight compounds and are calculated using group contribution methods, which could lead to solubility error prediction. Due to the lack of information on these properties, empirical models are often used for the correlation of experimental solubility data. These models are known as density-based models and consist of equations that contain constants that are empirically adjusted for each compound. Although simple, these models rely much on the knowledge of the thermodynamic behaviour of the supercritical solvent rather than of the solute, and are mostly capable of correlating rather than predicting the solubility. They are used for quantitative determination of the solute solubility in supercritical phase at equilibrium, and do not provide qualitative information about the solute-solvent interaction.

In a previous presentation [5], we have adopted a methodology for the correlation and the prediction of the solubility of 10 aromatic pollutants in the supercritical carbon dioxide. In this work we present an extension of the methodology to the solubility of other solids with different functional groups in different supercritical fluids namely carbon dioxide, ethylene and ethane. The methodology is based on the expanded liquid model theory [6,7] which does not require the knowledge of the solute critical properties and sublimation pressure. In this case the supercritical phase is considered as an expanded liquid and is modeled using excess Gibbs energy models such as Margules,

Van Laar, and local composition based models i.e. Wilson, NRTL and UNIQUAC. In this study we focus on the use of the UNIQUAC model that has been widely used in modeling vapour-liquid and liquid-liquid equilibria data. This model does not only take the size and nature of the molecules into consideration, but also accounts for the strength of solute-solvent intermolecular forces. And because the primary concentration variable is the surface fraction rather than mole fraction, the UNIQUAC model is applicable to solutions containing small or large molecules, including polymers. To assess the correlative and predictive capabilities of this model, a database is built by consisting of the experimental solubility data of 33 binary systems solid-super-critical fluid where solid solutes have different functional groups.

MODEL DEVELOPMENT

In the supercritical state, a fluid has a high density when compared with a gas. In fact, the density of a supercritical fluid is closer to that of a liquid than that of a gas. Consequently, in theoretical treatments the supercritical fluid phase can be treated approximately as an expanded liquid. This allows the phase equilibria between the solute and the supercritical fluid to be represented thermodynamically by solid-liquid equilibrium relations and conventional activity coefficients. To estimate the solid solubility in the supercritical phase, the knowledge of the activity coefficients are required. These coefficients are determined from the knowledge of the component fugacities, thus when the equilibrium of the pure solid and the supercritical phase is reached, we have:

$$f_2^s = f_2^{l,}$$

(1)

f_2^s is the fugacity of the solute in the solid phase considered as pure solid and equal to f_2^{os} because the solubility of supercritical fluid (SCF) in the solid phase is considered to be negligibly small. The f_2^l is the fugacity of the solid solute in the supercritical phase and is equal to:

$$f_2^L = \gamma_2 y_2 f_2^{oL} \tag{2}$$

Equation (1) could be written as follows:

$$f_2^{os} = \gamma_2 y_2 f_2^{oL} \tag{3}$$

where γ_2, y_2 and f_2^{oL} are the activity coefficient, the solid solubility represented in mole fraction and the fugacity of the pure solid solute in the expanded liquid phase respectively. According to Prausnitz et al. [8], we have:

$$\ln\left(\frac{f_2^{os}}{f_2^{ol}}\right) = \frac{-\Delta H_2^f}{R}\left(\frac{1}{T} - \frac{1}{T_m}\right) - \frac{\Delta C_p}{RT}\left(\frac{T - T_m}{T}\right)$$
$$+ \frac{\Delta C_p}{R}\ln\left(\frac{T}{T_m}\right) \tag{4}$$

Prausnitz et al. [8] stated that, to a fair approximation, the heat capacity terms can be neglected. Equations (3) and (4) then combined to yield an expression for the solute solubility:

$$y_2 = \frac{1}{\gamma_2}\exp\left[\frac{-\Delta H_2^f}{R}\left(\frac{1}{T} - \frac{1}{T_m}\right)\right] \tag{5}$$

ΔH_2^f is the enthalpy of fusion, T_m is the melting point temperature of the solid solute. Since the solid solubility in the supercritical phase is very small, we can assume to be at infinite dilution condition. Consequently, the activity coefficient of the solid solute is the one at infinite dilution and the density of the solution is that of the pure solvent. Thus Equation (5) becomes:

$$y_2 = \frac{1}{\gamma_2^\infty} \exp\left[\frac{-\Delta H_2^f}{R}\left(\frac{1}{T} - \frac{1}{T_m}\right)\right]$$

(6)

The activity coefficient of the solid solute at infinite dilution γ_2^∞ was calculated using the UNIQUAC model which consists of two parts, a combinatorial part $\gamma_2^{C,\infty}$ that attempts to describe the dominant entropic contribution, and a residual part $\gamma_2^{R,\infty}$ that is due primarily to intermolecular forces that are responsible for the enthalpy of mixing. The combinatorial part is determined only by the composition and by the sizes and shapes of the molecules; it requires only pure-component data. The residual part, however, depends also on intermolecular forces; the two adjustable binary parameters a_{12} and a_{21}, therefore, appear only in the residual part [8]:

$$\ln \gamma_2^\infty = \ln \gamma_2^{C,\infty} + \ln \gamma_2^{R,\infty}$$

(7)

$$\ln \gamma_2^{C,\infty} = 1 - \frac{r_2}{r_1} + \ln \frac{r_2}{r_1} - q_2 \frac{z}{2}\left(1 - \frac{r_2 q_1}{r_1 q_2} + \ln \frac{r_2 q_1}{r_1 q_2}\right)$$

(8)

Here q and r are the surface area and volume parameters; z is the coordination number that is usually taken equal to 10. The residual part at infinite dilution is given by the following equation [8]:

$$\ln \gamma_2^{R,\infty} = q_2 \left(1 - \ln \tau_{12} - \tau_{21}\right)$$

(9)

where

$$\tau_{12} = \exp\left(-\Delta u_{12}/RT\right) = \exp\left(-a_{12}/T\right)$$

(10a)

and

$$\tau_{21} = \exp\left(-\Delta u_{21}/RT\right) = \exp\left(-a_{21}/T\right) \tag{10b}$$

Δu_{12} and Δu_{21} are characteristic energies and are related to the interaction parameters a_{12} and a_{21} through Equation (10). Finally combining Equations (9) and (10) leads to:

$$\ln \gamma_2^{R,\infty} = q_2 \frac{a_{12}}{T} + q_2 \left(1 - e^{\frac{-a_{21}}{T}}\right) \tag{11}$$

Equation (11) could be written in reduced form by introducing the reduced temperature, thus we obtain:

$$\ln \gamma_2^{R,\infty} = q_2 \frac{a'_{12}}{T_r} + q_2 \left(1 - e^{\frac{-a'_{21}}{T_r}}\right) \tag{12}$$

With $a'_{12} = \dfrac{a_{12}}{T_c}$ and $a'_{21} = \dfrac{a_{21}}{T_c}$, T_c is the solvent critical temperature.

The binary interaction parameters a_{12} and a_{21} are related to the energy of interaction between the solid solute and the solvent in the supercritical phase, and cannot be kept constant and specifically at high pressure conditions. Therefore to take into account the pressure and temperature effects, these parameters are assumed to be density dependent and were fitted to the following equations:

$$a'_{12} = \alpha_{12} \cdot \rho_r^{\beta_{12}} \tag{13a}$$

$$a'_{21} = \alpha_{21} \cdot \rho_r^{\beta_{21}} \tag{13b}$$

ρ_r is the reduced density of the solvent equal to ρ/ρ_{c1} where ρ_{c1} is its critical density, a_{12}, β_{12}, a_{12} and β_{21} are the regressed parameters of the model.

DATABASE COMPILATION

By considering 33 system solid-SC fluid, an exhaustive solutes solubility database consisting of more than 2218 solubility data in supercritical fluids is built-up for the elaboration and validation of the proposed model. It is acknowledged that the systems studied do not include all the data available but should be sufficient to provide a thorough testing of the potential of Equations (6) to 13(b). The density of supercritical fluid solvents used in this work, is estimated using the Span and Wagner equation of state [10] when they are not reported in the solubility data sources, the physical properties of the solvents are given in Table 1. Table 2 show the solutes and their thermodynamic properties obtained from literature used in this study. Description of the database is given in Table 3 which lists the binary systems used together with the number of solubility data points, and the lower and upper limits of the operating conditions, solubility and the references. Detailed information about all the complete references from which experimental solubility data are taken are provided in Table 4.

SOLUTES SOLUBILITY CORRELATION

The surface area and volume parameters are calculated as the sum of the group volume and area parameters (R and Q) given by the UNIFAC group specifications [8]. These parameters and properties listed in Table 2, together with those of different fluids listed in Table 1 are used to calculate the combinatorial part of the activity coefficient from Equation (8). In other hand, Equation (12) is used to calculate the residual part of the solid solute activity coefficient. Thermodynamic properties of the solid solute listed in Table 2 are used together with Equations (7), (8), and (12) to estimate the solubility y_2 using Equation (6). The interaction parameters α'_{12} and α'_{21} are then regressed according to Equations 13(a) and 13(b). The best regression is based on minimizing the error between the regressed and experimental solubility data. The objective function used minimizes the sum of average absolute relative deviation (AARD) according to Equation (14):

$$FOBJ = AARD(\%) = \frac{100}{N} \sum_{1}^{N} \frac{\left| y_{2(exp)} - y_{2(regr)} \right|}{y_{2(exp)}}$$

(14)

where N is the number of experimental solubility data of each solute.

Table 1: Solvents physical properties

Solvent	Tc1 (K)	Pc1 (bar)	ω1	ρc1 (mol/ cm3) ×100	r1	q1
CO2	304.2a	73.83a	0.239a	1.063a	1.296a	1.261b
ethane	305.33a	48.8a	0.09a	0.687a	1.802a	1.696
ethylene	382.35a	50.4a	0.089a	0.764a	1.488a	1.574

aFrom reference [8], bFrom reference [9].

Table 2: Solute species properties

Solute	Formula	M (g/ mol)	Tm (K)	ΔH2f (J/ mol)	V2S (cm3 /mol)	r2	q2
anthracene	C14H10	178.23	492.5 [10]	28829.0 [11]	138.9 [11]	6.77	4.48
biphenyl	C12H10	154.21	342.1 (10]	18601.0 [11]	132.0 [11]	6.04	4.24
6- caprolactam	C6H11 NO	113.16	342.3 [12]	16096.0 [12]	162.0 [13]	4.67 [14]	3.736 [14]

1,10- decanediol	C10H22 O2	174.3	347.1 [15]	43505.3 [16]	158.4 [17]	8.74	7.8
2,3- dimethylnaphtalene	C12H12	156.23	377.8 [11]	25101.0 [11]	156.0 [11]	6.45	4.57
2,6- dimethylnaphtalene	C12H12	156.23	383.2 [11]	25055.0 [11]	199.0 [11]	6.45	4.57
2,7- dimethylnaphtalene	C12H12	156.23	369.0 [18]	23349.0 [18]	154.0 [18]	6.45	4.57
fluoranthene	C16H10	202.2	383.3 [15]	18728.0 [12]	163.0 [11]	7.5	4.72
fluorene	C13H10	116.23	387.9 [19]	19580.0 [19]	139.3 [19]	6.39	4.22
hexamethy lbenzene	C12H18	228.38	438.7 [19]	20640.0 [19]	152.7 [13]	7.59	5.8
lauric acid	C12H24 O2	200.3	316.9 [12]	36650.0 [12]	229.0 [20]	8.94	7.47
myristic acid	C14H28 O2	228.38	327.1 [15]	45362.8 [15]	257.5 [13]	10.29	8.55
naphtalene	C10H8	128.17	3532 [11]	19123.0 [11]	125.0 [11]	4.98	3.44
l-naphtol	C10H8O	144.2	368.1 [15]	23477.5 [15]	117.8 [21]	5.34	3.72
2-naphtol	C10H8O	144.2	396.0 [11]	17511.0 [11]	118.0 [11]	5.34	3.72
phenanthrene	C14H10	178.23	3722 [11]	16465.0 [11]	182.0 [11]	6.77	4.48
phenol	C6H6O	94.10	313.9 [11]	11289.0 [11]	89.0 [13]	3.55	2.68

Frew	C16H10	202.26	4242 [19]	17111.0 [19]	158.5 [19]	7.5	4.72
resorcinol	C6H6O2	110.11	383.1 [15]	21291.6 [15]	86.7 [21]	3.91	2.96
stearic acid	C18H36 O2	284.5	341.9 [15]	565692 [15]	302.4 [13]	12.99	10.71
triphenylene	C18H12	228.29	470.9 [15]	24190.0 [18]	175.0 [20]	8.57	5.52
triphenylmethane	C19H16	244.3	365.6 [19]	20920.0 [19]	240.9 [19]	9.512	6.588

SOLUTES SOLUBILITY PREDICTION

In order to evaluate the predictive ability of the proposed model, the solubility data for each considered component are split with specific tool in two sets. The first data set is the training set on which the minimization routine is performed and the interaction parameters are regressed. This data set contains 70% of the data randomly picked up from the experimental solubility data for each component. The second data set, namely the test set, contains the remaining 30% data and is intended for testing the generalized capabilities of the UNIQUAC model. The interaction parameters have been regressed using the training solubility data set, and then used directly to predict the solid solubility y_2 using the test data set. The predictive ability of the model is then assessed by comparing the obtained AARD values for each data set and so for each component.

Table 3: Database description

Binary systems	N	T-range (K)	P-range (bar)	ρ_r range	Solubility order	References
naphtalene-CO_2	242	308 - 338.05	75.4 - 400	0.51 - 1.98	7E-4 - 6E-2	L1 - L17
naphtalene-ethylene	166	285 - 323	54.5 - 303.9	0.52 - 1.99	3E-4 - 8E-2	L1, L6, L61, L62
naphtalene-ethane	48	308 - 328.2	48.1 - 250	0.56 - 1.99	8E-4 - 5E-2	L24, L59
anthracene-CO_2	206	298 - 343	92.6 - 470	0.66 - 2.0	3E-6 - 3E-4	L20, L23, L28, L29, L30, L31, L32, L33, L34
anthracene-ethane	13	308 - 343	104.3 - 345.6	1.49 - 1.99	7E-5 - 6E-4	L24
anthracene-ethylene	27	323 - 358	104.4 - 414.7	0.75-1.97	1E-5 - 9E-4	L59
biphenyl-CO_2	57	308 - 330.6	80 - 379.46	0.55 - 1.99	4E-4 - 6E-2	L9, L5, L12
biphenyl-ethane	8	308 - 318.3	70.5 - 180	1.32 - 1.94	1E-2 -4E-2	L24
6-caplolactam-CO_2	32	307 - 324	101 - 208	1.5 - 1.81	2E-2 - 7E-2	L46
2.3-dimethylnaphtalene-CO_2	25	308 - 328	99 - 280	0.68 - 1.96	3E-4 -9E-3	L21, L25
2.3-dimethylnaphtalene-ethylene	18	308 - 328	77 - 280	0.64 - 1.93	3E-4 - 5E-2	L21
2.6-dimethylnaphtalene-CO_2	23	308 - 328.2	79 - 280	0.63 - 1.96	3E-4 - 9E-3	L8, L21
2.6-dimethylnaphtalene-ethylene	18	308 - 328	78 - 280	0.62 - 1.94	2E-4 - 2E-3	L21
2.7-dimethylnaphtalene-CO_2	10	308.2 - 328.2	88 - 249	0.69 - 1.91	7E-4 - 1E-2	L8
fluoranthene-CO_2	68	308 - 348	86 - 354.6	0.62 - 1.99	9E-6 - 1E-3	L35, L26
fluorene-CO_2	146	308 - 343	80.9 - 483.4	0.6 - 1.98	6E-5 - 9E-3	L18, L19, L23
phenanthrene-CO_2	281	308 - 343	82.2 - 414.5	0.56 - 2.0	3E-5 - 4E-3	L12, L17, L18, L19, L20, L21, L22, L24, L25, L26, L63
phenanthrene-ethylene	42	298 - 343	69.9 - 280	0.45 - 1.49	3E-5 - 1E-2	L59, L21, L19
phenanthrene-ethane	17	313 - 333	69.9 - 264	0.76 - 1.95	2E-4 - 6E-3	L19
pyrene-CO_2	235	308 - 343	83.4 - 483.4	0.78 - 1.99	1E-5 - 9E-4	L18, L19, L20, L27
pyrene-ethane	15	333 - 333	100.2 - 314.5	1.37 - 1.98	2E-4 - 1E-3	L20
triphenylene-CO_2	53	308 - 328	85 - 355.6	0.83 - 1.99	1E-6 - 5E-5	L35, L27
triphenylmethane-CO_2	111	308 - 328	73.7 - 414.5	0.507 - 1.99	3E-5 - 4E-3	L19, L38
lauric acid-CO_2	24	308 - 318	77 - 260	0.53 - 1.94	2E-5 - 5E-2	L43, L44, L45
palmitic acid-CO_2	27	308 - 328	91 - 248	1.1 - 1.9	3E-5 - 2E-3	L4, L52, L53
myristic acid-CO_2	30	308 - 318	128.5 - 226.5	1.47 - 1.89	1E-3 - 1E-2	L43, L44
stearic acid-CO_2	28	308 - 338	90 - 237	0.94 - 1.91	2E-5 - 1E-3	L56, L60, L57, L58
Resorcinol	26	308 - 338	121.6 - 405.3	0.84 - 1.97	1E-4 -9E-4	L55
phenol-CO_2	73	309 - 333.2	79.3 - 249.43	0.51 - 1.91	1E-3 - 6E-3	L40, L41,
1,10-decanediol	15	318 - 328	133.7 - 307.3	1.26 - 1.91	3E-4 - 8E-4	L47
1-naphtol-CO_2	64	308 - 328	88.7 - 296.5	0.54 - 1.98	6E-5 - 2E-3	L33, L51
2-naphtol-CO_2	48	308.05 - 343.2	100 - 363.6	0.65 - 1.99	8E-5 - 2E-3	L24, L37, L29
2-naphtol-ethane	22	308 - 343.2	61 - 364	0.56 - 1.98	2E-5 - 1E-3	L24

Table 4: References for solubility data

L1: Russian journal of physical chemistry 1964; 38: 9	L30: Ind. Eng. Chem. Res. 1987, 26, 7, 1476-1482
L2: J. Supercritical Fluids. 1996; 9: 3	L31: J. Chem. Eng. Data 1996, 41, 97-100
L3: J. Chem. Eng. Data 1998; 43: 400-402	L32: J. Chem. Eng. Data 1997, 42, 636-640
L4: J. Chem. Eng. Data 1999; 48: 951-957	L33: J. Chem. Eng. Data 1995, 40, 953-858
L5: J. Chem. Eng. Data 1980; 25, 4, 326-329	L34: Utilisation des fluides supercritiques pour l'extraction des fullerènes, thesis presented by Valérie Quillet,1996: Bordeaux 1, France
L6: J. Supercritical. Fluids 1988; 1: 1	L35: J. Chem. Eng. Data 2000, 45. 53
L7: J. Chem. Eng. Data 1989, 36. 4. 430-432	L37: Ind. Eng. Chem. Res 26, 1, 1987, 56-65
L8: J. Chem. Eng. Data 1993, 38, 3,	L38: J. Supercritical Fluids 2004, 32, 115-121
L9: Fluid Phase Equilibria 1992; 81:321-341	L39: Fluid Phase Equilibria 2004, 226. 9-13
L10: Fluid Phase Equilibria 1995, 107, 189-200	L40: J. Chem. Eng. Data 1980, 25, 257-259
L11: Ind. Eng. Chem. Res. 2000, 39, 4609-4614	L41: Hwahak Konghak Journal 1993, 31, 6, 637
L12: J. Supercritical. Fluids 1995, 8, 1, 15-19	L43: J. Chem. Eng. Data 1988, 33, 3, 327-333
L13: J. chem. Eng. Data 1988, 33, 1, 35-37	L44: J. Chem. Eng. Data 2008, 53, vol. 11
L14: J. Chem. Eng. Data 1985, 30, 1	L45: J. Am. Oil. Chem. Soc. 1992, 69, p. 1069
L15: J. Physical Chemistry 1986, 90, 17	L46: J. Chem. Eng. Data 1996, 41, 1418-1420
L16: J. Chem. Eng. Data 2000, 45: 464-466	L47: J. Chem. Eng. Data 1986, 31, 285-288
L17: J. Chem. Eng. Data 2000, 45: 358-361	L51: Fluid Phase Equilibria 1987, 34. 37-47
L18: J. Chem. Eng. Data 1990, 35: 355-360	L52: J. Chem. Eng. Data 1991, 36, 4, 430-432
L19: Ind. Eng. Chem. Fundamentals 1982, 21, 3	L53: J. Chem. Eng. Data 1988, 33, 230-234
L20: J. Supercritical Fluids 1997, 10, 175-189	L55: Fluid Phase Equilibria 1998, 152, 299-305
L21: J. Chem. Eng. Data 1981, 26, 1, 47-51	L56: J. Chem. Eng. Data 1993, 38, 506-508
L22: Ber. Bunsen. Phys. Chem. 1984. 88, 865-869	L57: J. Chem. Eng. Data 2008, 53, 2913-2917
L23: Phys. Chem. 1984, 88, 865-869	L58: J. Chem. Thermodynamics 2010, 42, 193-197
L24: J. Chem. Eng. Data 1986, 31, 2, 204-212	L59: AIChE journal 1981, 27, No 5, 773-779
L25: J. Chem. Eng. Data 2001, 46, 5, 1156-1159	L60: J. Chem. Eng. Data 1989, 34, 184-187
L26: J. Chem. Eng. Data 1996, 41, 1466-1469	L61: J. Amer. Chemic. Socie 1953, 57, 575-578
L27: Ind. Eng. Chem. Res 1995, 34, 340-346	L62: J. Amer. Chemic. Socie 1948,70, 4085-4089
L28: J. Chem. Eng. Data 1987, 32, 148-150	L63: J. Chem. Eng. Data 1986, 31, 3, 303-308
L29: Fluid Phase Equilibria 2003, 207, 183-192	

RESULTS AND DISCUSSION

Solubility Data Correlation

The analysis of the correlation model results is done through statistical calculations. Table 5provides the quantitative results of the regression for the UNIQUAC model. The AARD is included for each binary system and is listed together with the adjustable parameters values. Each component parameters are obtained by fitting them on its own solubility data available in the database. For all investigated systems the resultant AARD values are lower than 20% (except for the system phenol-CO_2). The model achieves an overall AARD of 12.16 on a range of 2.8 to 29.7. These results are indicative of a good correlation performance of the proposed model and also show that the calculated solubility data are in good agreement with the experimental ones.

In order to compare more clearly, the experimental data and correlated data in this work are compared in Figures 1-3 with naphthalene in SC CO_2 at T = 308 K, naphthalene in SC ethylene at T = 298 K, and naphthalene in SC ethane at T = 308 K respectively.

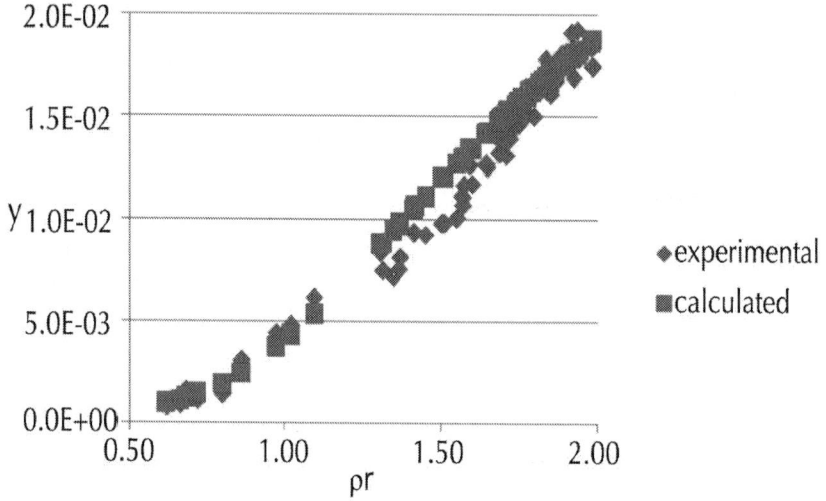

Figure 1: Experimental and correlated solubility versus reduced density of naphthalene-CO_2 system at 308 K.

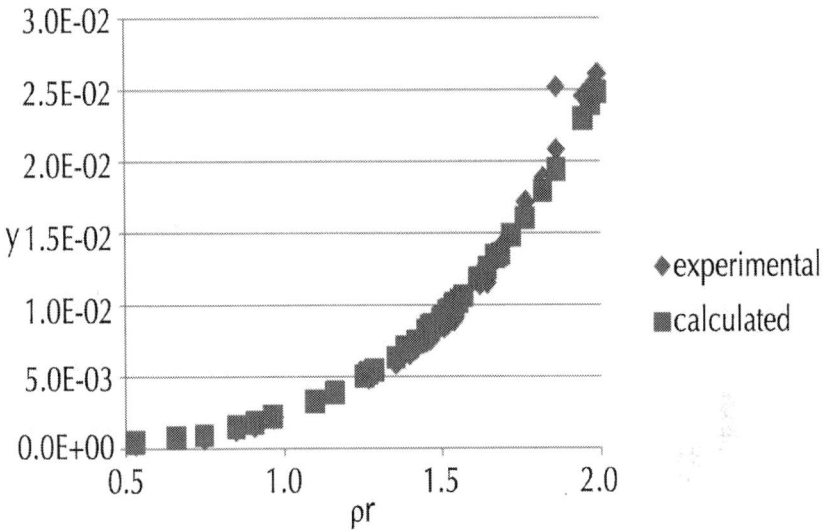

Figure 2: Experimental and correlated solubility versus reduced density of naphthalene-ethylene system at 298 K.

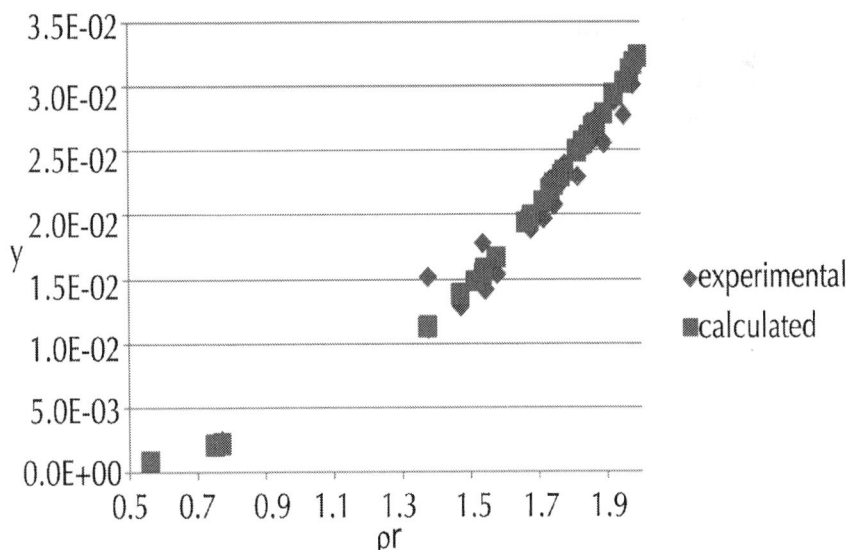

Figure 3: Experimental and correlated solubility versus reduced density of naphthalene-ethane system at 308 K.

These systems are arbitrarily taken from the database for illustration. From the figures, we see that it is good accordance with experimental and calculated ones, and the precision of the model in this work is very acceptable.

Comparison with Literature

We have considered for the comparison some literature models that account for effect of the system conditions (temperature and pressure) in addition to the physical properties as sublimation pressure of the solute through their introduction of the enhancement factor and a model based on a modified equation of state.

Wang and Tavlarides Model

Based on a dilute solution theory, Wang and Tavlarides [22] proposed the following model:

$$\frac{1}{T\left[\ln(E)-\ln(Z)\right]} = R\left(\frac{c}{b}+\frac{V_1}{b}\right), \quad \left(V_1 = \frac{1}{\rho_1}\right)$$

(15)

where Z is the compressibility factor, E is the enhancement factor that is defined as the enhancement of actual mole fraction solubility of the solid solute y_2 over the solubility in an ideal gas P_2^S/P, i.e.,

$$y_2 = \frac{EP_2^S}{P}, \quad P_2^S$$

and $\overset{.}{P}$ are the sublimation vapor pressure and total pressure respectively. By introducing reduced variables, the dimensionless form of Equation (15) is given by:

$$\frac{1}{T_r\left[\ln(E)-\ln(Z)\right]} = \alpha\left(\beta+\frac{1}{\rho_r}\right)$$

(16)

$$\alpha = \frac{T_{c_1} R}{\rho_{c_1} b}$$

where   ; $\beta = c\rho_{c_1}$.

Méndez-Santiago and Teja Models

Based on the theory of dilute solutions, Méndez-Santiago and Teja [23] began with the Henry's constant and presented a simple linear relationship for the solubility of solids in supercritical fluids as given below:

$$T \ln E = A_1 + B_1 \rho \tag{17}$$

To apply the model to compounds whose sublimation vapor pressure is unknown, a Clausius-Clapeyron-type expression for the

sublimation pressure $\ln P_2^s = A + \dfrac{B}{T}$ and $E = \dfrac{y_2 P_2^s}{P_i}$ are substituted into Equation (17) resulting in the following expression:

$$T \ln(y_2 P) = A' + B'\rho + C'T \tag{18}$$

The dimensionless form of Equations (17) and (18) are given bellow [4]:

$$T_r \ln E = T_r \ln \frac{y_2 P}{P_2^s} = \alpha + \beta \rho_r \tag{19}$$

where:

$$\alpha = \frac{A_1}{T_{c_1}} \quad \beta = \frac{B_1 \rho_{c_1}}{T_{c_1}}$$

$$T_r \ln\left(y_2 P_r\right) = \alpha' + \beta' \rho_r + \gamma' T_r$$

(20)

where:

$$P_r = \frac{P}{P_{c_1}} \quad ; \quad \alpha' = \frac{A'}{T_{c_1}} \quad ; \quad \beta' = \frac{B' \rho_{c_1}}{T_{c_1}} \quad ; \quad \gamma' = C' - T_r \ln P_{c_1}$$

The linear behavior of Equation (17) is well observed in the range (0.5 - 2) of reduced density of the super- critical solvent [23], for this reason we have omitted from the database all data points over this range of density (see Table 3).

Table 5: Regression parameters and average deviation of the model

Binary systems	N	α_{12}	β_{12}	α_{21}	β_{21}	AARD (%)
anthracene-CO_2	206	2.380	−0.210	−0.360	0.598	14.58
anthracene-ethylene	27	2.284	−0.356	−0.277	0.841	8.06
anthracene-ethane	13	1.840	−0.711	−0.053	0.844	6.87
biphenyl-CO_2	57	1.620	−0.450	−0.107	0.930	16.39
biphenyl-ethane	8	1.303	−0.403	−0.066	1.942	8.99
6-caprolactam	32	1.630	0.851	−0.330	1.940	5.66
1,10-decanediol-CO_2	15	1.803	−0.357	−0.090	−0.370	3.94
2,3-dimethylnaphtalene-CO_2	25	0.680	−0.370	2.424	−2.160	19.71
2,3-dimethylnaphtalene-ethylene	18	0.525	−0.827	3.659	−2.390	13.04
2,6-dimethylnaphtalene-CO_2	23	0.900	−0.560	0.802	−0.702	9.68
2,6-dimethylnaphtalene-ethylene	18	0.513	−0.896	3.240	−2.440	15.97
2,7-dimethylnaphtalene-CO_2	10	1.192	−0.332	0.271	−1.579	2.78
fluoranthene-CO_2	68	1.310	−0.150	1.209	−2.010	12.55
fluorene-CO_2	146	1.040	−0.370	1.174	−1.300	6.87
lauric acid-CO_2	24	0.680	−0.600	2.480	−2.090	11.48
myristic acid-CO_2	30	1.260	−0.470	0.377	−2.150	15.42
naphthalene-CO_2	242	0.920	−0.090	0.757	−2.450	16.57
naphthalene-ethylene	166	0.741	−0.490	1.327	−1.779	19.26

naphthalene-ethane	48	0.718	−0.705	0.675	−1.194	12.01
1-naphtol-CO₂	64	1.350	−0.280	2.695	−1.960	16.05
2-naphtol-CO₂	48	1.530	−0.114	1.735	−2.260	12.31
2-naphtol-ethane	22	1.660	−0.115	1.550	−1.916	17.46
palmitic acid-CO₂	27	1.530	−0.380	6.295	−42.14	10.42
phenanthrene-CO₂	281	1.970	−0.350	0.053	1.950	13.15
phenanthrene-ethylene	42	2.060	−0.370	−0.120	1.810	16.31
phenanthrene-ethane	17	2.040	−0.180	−0.230	1.470	14.67
phenol-CO₂	73	1.460	−0.200	0.990	−1.530	29.70
pyrene-CO₂	235	2.190	−0.370	−0.061	1.890	4.78
pyrene-ethane	15	1.870	−0.460	−0.060	1.750	1.16
resorcinol-CO₂	26	3.120	−0.290	−0.020	3.250	6.71
stearic acid-CO₂	28	1.120	−0.320	0.461	−0.500	8.49
triphenylene-CO₂	53	3.370	−0.300	−0.800	−1E-4	4.90
triphenylmethane-CO₂	111	1.800	−0.170	−0.270	0.690	15.08

Schmitt and Reid Model

By the assumption that: the system pressure is much greater than the sublimation pressure of the solute, that the solute is incompressible, and that no solvent is dissolving in the solid phase, the solubilities of solids in supercritical solvents are usually correlated with the equation bellow [24]:

$$
y_2 = \left(\frac{P_2^S}{P} \right) \frac{\exp\left(\dfrac{v_2^S}{RT} \left(P - P_2^S \right) \right)}{\Phi_2^{sf}}
$$

(21)

V_2^S is the molar volume of the solute (Table 2), the fugacity coefficient of the solute in the supercritical phase Φ_2^{sf} is determined from an equation of state applicable to the solute-solvent mixture. The Peng-Robinson equation of state, i.e. PR-EOS is a commonly used approach for correlating solubility in supercritical fluids.

Schmitt and Reid [24] used this equation of state for modeling solid solubilities in supercritical fluids but not in the traditional manner due to the lack of the critical properties values and the low accuracy of their estimation. In fact they eliminated the binary interaction parameter and they assumed the solid solute parameters "a_2" and "b_2" independent of temperature and they eliminated the terms with y_2 in the combining and mixing rules since the values of y_2 were sufficiently small. Thus they propose the simplified equation bellow for the fugacity coefficient:

$$\ln \Phi_2^{sf} = (b_2/b_1)(Z-1) - \ln\left[P(V-b_1)/RT\right]$$
$$-\left(a_1/8^{0.5} RTb_1\right)\left[2(a_2/a_1)^{0.5} - (b_2/b_1)\right]$$
$$\times \ln\left[(V+2.414b_1)/(V-\sqrt{2}b_1)\right]$$

$$(22)$$

For the supercritical solvent, the parameters a_1 and b_1 are given by:

$$a_1 = 0.4572R^2 T_{c_1}^2/P_{c_1}$$
$$\times\left[1+\left(0.3746+1.5423\omega_1-0.2699\omega_1^2\right)\left(1-\left(T/T_{c_1}\right)^{0.5}\right)\right]^2$$

$$b_1 = 0.0778RT_{c_1}/P_{c_1}$$

Equation (21) becomes,

$$\ln y_2 = \ln\left[P_2^s/P\right] + \frac{\left(P-P_2^s\right)v_2^s}{RT}$$
$$-\left[(b_2/b_1)(Z-1) - \ln\left[P(V-b_1)/RT\right]\right.$$
$$-\left(a_1/8^{0.5} RTb_1\right)\left[2(a_2/a_1)^{0.5} - (b_2/b_1)\right]$$
$$\left.\times \ln\left[(V+2.414b_1)/(V-\sqrt{2}b_1)\right]\right]$$

$$(23)$$

The experimental solubility data are regressed by using Equation (23) to determine the parameters of the solute a_2 and b_2 for each binary system considered. These parameters are expressed in Pa (m³/mol)² and m³/mol respectively. The Equation (23) is dimensionally consistent and already dimensionless, but the parameters "a_2" and "b_2" are not. For comparison on the same basis the dimensionless parameters are as follow:

$$a_2' = \left(a_2 \rho_{c_1}^2 / P_{c_1} \right) \quad \text{and} \quad b_2' = \left(b_1 \rho_{c_1} \right)$$

Analysis and Discussion

As mentioned in the previous section, the three first correlations used for comparison require the knowledge of the sublimation pressure. Table 6 gives the coefficients for the estimation of the sublimation pressure of the different solutes taken from literature and the temperature range of applicability.

To compare all the correlations on the same basis, the average absolute relative deviation (AARD) is determined for each system's solute using each model. Table 7 display the parameters of the comparison models and shows in the last column the AARD produced by each equation for the different fluid-solute systems considered. The lowest error fit is marked with (*), the proposed model gives the lowest AARD for 11 binary systems and provides a better fit than the modified EOS for 24 binary systems. From this comparison we want to emphasize that the proposed UNIQUAC model provides better quantitative capabilities with one common EOS which does not require the knowledge of solute critical properties.

Judging from the mean AARD values given in Table 8, we can see clearly that the UNIQUAC model gives good results compared to the other models. The best fit is achieved by the Mendez-Teja model with three parameters given by Equation (20). Whereas, the largest values of AARD and the highest mean AARD are achieved by the model of Wang-Tavlarides (Equation (16)). However, as acknowledged by the authors [22], this model somewhat oversimplified a real system since the interaction between solute and solvent molecules is assumed to follow an interaction behavior similar to the potential well model. In this theory, the system consists of a free volume and a constant solvent cluster volume (i.e., they assume that this volume does not vary with temperature and pressures); the solute thus is either a quasi-gas type (moving in the cluster) or an ideal gas type. As a result, the model developed produces a large error for the supercritical fluid solubility.

Solubility Prediction

Binary Systems

In order to evaluate the predictive capabilities of the UNIQUAC model and to overcome the over-fitting problem which may alter the model generalization capabilities, solubility data for each considered component were split into two sets.

Table 6: Coefficients for sublimation pressure estimation

Solute	$P^s (Pa) = 10^{\left(A - \frac{B}{T}\right)}$			
	A	B	T-range (K)	Reference
anthracene	14.755	5313.7	308 - 343	[24]
biphenyl	14.804	4367.4	308 - 343	[24]
6-caprolactam	15.480	4811.1	307 - 324	[25][a]
1,10-decanediol	20.901	7217.0	308 - 328	[26][a]
2,3-dimethylnaphtalene	14.065	4302.5	308 - 328	[27]
2,6-dimethylnaphtalene	14.417	4415.9	308 - 328	[27]
2,7-dimethylnaphtalene	14.464	4386.7	308 - 328	[28]
fluoranthene	14.793	5357.0	298 - 358	[29]
fluorene	14.205	4561.8	308 - 343	[27]
lauric acid	22.022	7322.0	295 - 314	[15]
myristic acid	20.861	7291.0	311 - 325	[15]
naphthalene	13.865	3823.1	250 - 340	[15][a]
1-naphtol	10.683	3148.9	308 - 328	[24]
2-naphtol	14.815	4923.9	308 - 343	[24]
palmitic acid	22.341	8069.0	308 - 328	[20]
phenanthrene	14.343	4776.7	300 - 360	[15][a]
phenol	13.689	3586.4	309 - 333	[20]
pyrene	13.395	4904.0	308 - 398	[30]
stearic acid	21.021	7957.0	308 - 338	[31]
triphenylene	14.462	5804.1	300 - 340	[20]
triphenylmethane	14.781	5228.0	308 - 328	[19]

[a]Data interpolated in this work.

[a]Data interpolated in this work.

The first set of solubility data obtained from randomly sampling 70% of the experimental data, served for the optimization of the model adjustable parameters and for training the model. The second set of solubility data was then used to test its predictive capabilities. In this step we have considered only solutes for which solubility data points are more than 20. Table 9summarizes the key information on the two data sets, namely, the training and the test data sets together with the correlation and prediction results. This table lists the number of solubility data points used for each component and deviations in term of the AARD values obtained for both the training and the test data sets. To assess the predictive ability of the model, the two AARD values are compared. To be predictive, the model should respect the following rule: the AARD values for both data sets should be of the same order of magnitude for each component. Table 9 shows that the AARD values obtained for the test data set are in accordance with those of the training data set and are generally of the same order of magnitude. This results implies that the proposed model do not show any over-fitting problem or over prediction of the experimental solubility data. Therefore we can conclude that the predictive ability of the model is well demonstrated.

Ternary Systems

The study of mixed-solute systems is important because most potential applications of supercritical fluid extraction involve the removal of a desired compound from a matrix of components. However, in this section an attempt is made to predict the solubilities of mixed compounds in supercritical carbon dioxide. Experimental data provided by: Kosal and Holder [32] for mixed an thracene and phenanthrene, Pennisi and Chimowitz [26] for mixed 1,10-decanediol and benzoic acid, Iwai el al. [33] for mixed 2,6- and 2,7-dimethylnaphtalenes are used.

Table 7: Comparison with the literature models considered

System	N	α	β	α'	β'	γ'	a'_2	b'_2	AARD%
anthracene-CO_2	206								
Equation (19)		4.10	4.73						14.88
Equation (20)				−33.09	4.58	15.57			14.41[*]
Equation (16)		16.17	0.65						21.44
Equation (23)							8.7E-5	1.24E-4	16.74
UNIQUAC model									14.58
anthracene-ethylene	27								
Equation (19)		6.33	5.22						14.30
Equation (20)				−32.12	5.33	12.80			7.60[*]
Equation (16)		18.05	0.776						21.80
Equation (23)							1.26E-4	1.56E-4	17.05
UNIQUAC model									8.06
anthracene-ethane	13								
Equation (19)		3.42	5.05						10.38
Equation (20)				−33.06	4.77	15.76			5.66[*]
Equation (16)		15.00	0.602						22.64
Equation (23)							1.19E-4	1.54E-4	10.96
UNIQUAC model									6.87
biphenyl-CO_2	57								
Equation (19)		3.87	4.04						17.69
Equation (20)				−35.57	4.19	24.15			14.48
Equation (16)		17.58	0.910						13.11[*]
Equation (23)							8.39E-5	1.4E-4	19.14
UNIQUAC model									16.39
biphenyl-ethane	8								
Equation (19)		3.11	4.42						9.81
Equation (20)				−40.01	4.97	27.71			5.76
Equation (16)		15.06	0.705						3.77[*]
Equation (23)							1.0E-4	1.41E-4	6.59
UNIQUAC model									8.99
6-caprolactam-CO_2	32								
Equation (19)		−0.08	7.96						6.95
Equation (20)				−29.20	6.89	14.50			7.51
Equation (16)		9.71	0.08						12.71
Equation (23)							6.36E-5	4.33E-5	16.40
UNIQUAC model									5.66[*]

	N								AARD
1,10-decanediol-CO$_2$	15								
Equation (19)		5.40	4.55						4.22
Equation (20)				−45.09	4.37	28.71			3.03*
Equation (16)		20.29	0.859						6.60
Equation (23)							1.08E-4	1.47E-4	6.07
UNIQUAC model									3.94
2,3-dimethylnaphtalène-CO$_2$	25								
Equation (19)		2.68	4.74						20.05
Equation (20)				−27.72	4.56	14.76			19.74
Equation (16)		12.57	0.493						26.07
Equation (23)							7.19E-5	1.14E-4	32.97
UNIQUAC model									19.71*
2,3-dimethylnaphtalene-ethylene	18								
Equation (19)		2.82	5.72						16.78
Equation (20)				−42.37	6.48	24.85			14.62
Equation (16)		12.67	0.373						16.37
Equation (23)							9.26E-5	1.10E-4	17.80
UNIQUAC model									13.04*
2,6-dimethylnaphtalène-CO$_2$	23								
Equation (19)		4.23	3.77						12.75
Equation (20)				−31.54	3.90	19.42			11.78
Equation (16)		17.16	0.885						11.85
Equation (23)							8.66E-5	1.57E-4	13.56
UNIQUAC model									9.68*
2,6-dimethylnaphtalene-ethylene	18								
Equation (19)		2.73	5.59						11.77
Equation (20)				−38.84	6.13	21.93			10.31*
Equation (16)		14.28	0.521						13.81
Equation (23)							8.64E-5	1.12E-4	10.41
UNIQUAC model									15.97
2,7-dimethylnaphtalène-CO$_2$	10								
Equation (19)		4.87	3.44						8.52
Equation (20)				−30.31	3.63	19.05			6.73
Equation (16)		18.66	1.004						9.95
Equation (23)							9.79E-5	1.74E-4	12.62
UNIQUAC model									2.78*
fluoranthene-CO$_2$	68								

Equation (19)		6.73	4.42						18.38
Equation (20)				−29.06	4.28	13.99			11.21*
Equation (16)		24.59	1.06						35.14
Equation (23)							1.49E-4	1.96E-4	24.30
UNIQUAC model									12.35
fluorene-CO_2	146								
Equation (19)		4.74	3.97						7.59
Equation (20)				−29.77	3.97	16.87			7.59
Equation (16)		19.63	0.99						15.23
Equation (23)							9.59E-5	1.53E4	13.80
UNIQUAC model									6.87*
lauric acid-CO_2	24								
Equation (19)		4.39	6.83						13.56
Equation (20)				−37.51	6.66	21.85			11.06*
Equation (16)		19.63	0.55						14.77
Equation (23)							1.39E-4	1.54E-4	14.66
UNIQUAC model									11.48
myristic acid-CO_2	30								
Equation (19)		7.41	5.72						14.66
Equation (20)				−48.06	5.72	32.48			14.65*
Equation (16)		25.65	0.82						17.20
Equation (23)							1.88E-4	2.13E-4	15.12
UNIQUAC model									15.45
naphtalene-CO_2	242								
Equation (19)		3.39	3.31						19.45
Equation (20)				−34.37	3.68	24.12			11.79
Equation (16)		15.55	0.948						11.13*
Equation (23)							6.77E-5	1.34E-4	14.37
UNIQUAC model									16.57
naphthalene-ethylene	167								
Equation (19)		3.24	4.18						16.02
Equation (20)				−32.94	4.54	20.65			1.28
Equation (16)		14.00	0.655						10.53*
Equation (23)							8.43E-5	1.37E-4	16.46
UNIQUAC model									19.26
naphthalene-ethane	48								
Equation (19)		2.65	3.70						15.19

Equation (20)				−32.25	3.85	22.23			12.47
Equation (16)		14.22	0.833						10.15[*]
Equation (23)							8.03E-5	1.37E-4	14.12
UNIQUAC model									12.01
1-naphtol-CO$_2$	64								
Equation (19)		4.02	3.83						20.41
Equation (20)				−35.10	3.86	21.15			17.99
Equation (16)		18.48	1.015						18.27
Equation (23)							7.32E-5	1.18E-4	23.33
UNIQUAC model									16.05[*]
2-naphtol-CO$_2$	48								
Equation (19)		3.71	4.53						14.08
Equation (20)				−30.50	4.30	15.73			11.91[*]
Equation (16)		15.55	0.948						84.51
Equation (23)							8.28E-5	1.24E-4	18.93
UNIQUAC model									12.31
2-naphtol-ethane	22								
Equation (19)		2.68	4.13						8.74
Equation (20)				−33.93	4.10	18.27			8.68[*]
Equation (16)		13.55	0.704						16.68
Equation (23)							8.4E-5	1.26E-4	10.92
UNIQUAC model									17.46
Palmitic acid-CO$_2$	27								
Equation (19)		8.26	5.34						7.85
Equation (20)				−52.37	5.32	35.22			7.70[*]
Equation (16)		26.65	0.86						14.10
Equation (23)							2.03E-4	2.34E-4	13.15
UNIQUAC model									10.42
phenanthrene-CO$_2$	281								
Equation (19)		4.92	4.37						13.57
Equation (20)				−32.20	4.40	18.06			13.37
Equation (16)		20.59	0.96						17.34
Equation (23)							1.07E-4	1.64E-4	17.27
UNIQUAC model									13.15[*]
phenanthrene-ethylene	42								
Equation (19)		4.08	5.83						12.55
Equation (20)				−31.68	5.64	15.03			9.55[*]

Equation (16)		16.54	0.55						21.72
Equation (23)							1.35E-4	1.66E-4	16.80
UNIQUAC model									16.31
phenanthrene-ethane	17								
Equation (19)		2.21	5.81						16.13
Equation (20)				−30.77	5.83	14.73			15.11
Equation (16)		13.80	0.49						19.45
Equation (23)							1.09E-4	1.31E-4	17.85
UNIQUAC model									14.67*
phenol-CO$_2$	73								
Equation (19)		2.12	3.23						28.34
Equation (20)				−36.15	3.65	25.92			24.84*
Equation (16)		12.32	0.82						26.54
Equation (23)							4.66E-5	1.01E-4	26.38
UNIQUAC model									29.70
pyrene-CO$_2$	235								
Equation (19)		5.66	4.89						9.60
Equation (20)				−31.67	4.79	14.92			6.41
Equation (16)		20.75	0.85						14.44
Equation (23)							1.14E-4	1.52E-4	11.24
UNIQUAC model									4.78*
pyrene-ethane	15								
Equation (19)		5.34	4.95						2.89
Equation (20)				−30.24	4.95	13.83			2.89
Equation (16)		20.55	0.87						1.99
Equation (23)							1.49E-4	1.7E-4	6.07
UNIQUAC model									1.16*
resorcinol-CO$_2$	26								
Equation (19)		2.83	4.38						6.02
Equation (20)				−33.73	4.42	17.67			5.05*
Equation (16)		11.67	0.61						66.21
Equation (23)							4.58E-5	7.38E-5	65.60
UNIQUAC model									6.71
stearic acid-CO$_2$	28								
Equation (19)		10.55	4.59						8.80
Equation (20)				−53.95	5.16	35.78			7.53*
Equation (16)		24.25	0.65						30.96

Equation (23)							1.98E-4	2.08E-4	31.30
UNIQUAC model									8.49
triphenylene-CO_2	53								
Equation (19)		8.15	5.04						12.84
Equation (20)				−32.91	5.03	17.78			4.77*
Equation (16)		26.18	1.01						20.40
Equation (23)							1.59E-4	1.92E-4	14.19
UNIQUAC model									4.90
triphenylmethane-CO_2	111								
Equation (19)		7.46	4.42						13.15
Equation (20)				−32.29	4.42	22.69			13.17
Equation (16)		35.63	1.68						10.90*
Equation (23)							1.65E-4	2.22E-4	24.71
UNIQUAC model									15.09

Table 8: AARD range and mean AARD

	Solvent CO2		Solvent ethane		Solvent ethylene		Mean AARD (solvent CO2)	Mean AARD (solvent ethylene)	Mean AMU) (solvent ethane)
	minima	maxima	minima	maxima	minima	maxima			
Equation (19)	4.20	28.34	2.9	16.3	11.7	16.8	13.34	14.3	10.5
Equation (20)	3.03	24.84	2.9	15.1	7.6	14.6	11.21	10.7	8.40
Equation (16)	6.60	84.51	1.99	22.6	10.5	21.8	22.68	16.8	12.4
Equation (23)	6.07	65.60	6.07	17.85	10.4	17.8	2027	15.7	11.1
UNIQUAC model	2.78	29.70	1.16	17.46	8.06	19.2	11.70	14.5	10.19

The solubility data of the considered compounds are very small and have an order of 10^{-6} - 10^{-3}. As a consequence, we can assume that the density of the supercritical phase is that of the pure solvent, and the activity coefficient of each solute is the one at infinite dilution. In this case the only interaction parameters that are taken into account are those of solutes-CO_2. Therefore predicted solubilities are estimated using Equations (6) to 13(b) and interaction parameters listed in Table 5 are directly implemented to estimate the solubility of each component in the mixture.

The interpolation results are given in Table 10, the absolute average relative deviation (AARD) obtained are generally low confirming the predictive ability of the model but have an order of magnitude considerably higher than those for the solutes in binary systems. This can be attributed to the fact that effect of solute-solute interactions is not always negligible and assuming an infinite dilution of all solutes can be inaccurate in some systems. However, there is significant experimental evidence to suggest that solute-solute interactions are important even in dilute solutions [34-36].

CONCLUSIONS

In this work, we have proposed the correlation and prediction of the solubility of solid solutes with different functional groups in chemically diverse supercritical fluids with a methodology based on the expanded liquid theory, in which the solid-fluid equilibrium is modeled using the local composition model of UNIQUAC. For the elaboration of the model we have considered 33 systems solid-SC fluid to built-up an exhaustive solutes solubility database consisting of more than 2218 experimental solubility data in supercritical fluids taken from literature. The proposed model achieves an overall AARD of 11.85 on a whole database and on a range of 1.16 to 29.7, these results show that the calculated solubility data are in good agreement with the experimental ones.

Table 9: Prediction capabilities of the model

Systhem	Training set		Test set	
	N	AARD%	N	AARD%
andracene-CO2	144	14.51	62	14.70
anthracene-ethy tem	18	S.36	9	7.57
biphenyl-0O2	39	15.15	18	20.27
6-eaptolactam-CO2	22	6.18	10	4.76
2,3-dunethy Inaphlakne-CO2	17	16.93	8	21.42
2,6-damn/l) hraphtakne-CO2	16	8.10	7	15.24
flumanthene-CO2	47	11.33	20	16.22
fluorcne-CO2	102	6.48	44	7.86
lauric acid-CO2	16	10.06	8	20.48
myristic acid-CO2	20	12.98	10	18.79
naphthalene-CO2	169	16.36	73	17.28
naphthalene-ethy lene	116	1850	50	21.20
naphthalene-ethane	33	10.16	15	16.42
1-naphtol-CO2	44	15.76	20	17.05
2-naphtol-CO2	33	8.32	IS	21.22
2-naphtol-ethane	15	18.24	7	17.20
palmitic acid-CO2	18	10.54	9	1223
phenanthrene-CO2	196	13.44	85	1154
phenanthrene-ethylene	29	16.90	13	15.68
phenol-CO2	51	26.18	22	47.01

pyrene-CO2	164	4.71	71	5.04
Resorcinol-CO2	18	6.51	8	7.17
stearic acid-CO2	19	7.87	9	11.15
triphenylene-CO2	37	5.05	16	5.66
triphenylmethane-CO2	77	16.73	34	11.90

Table 10: Prediction results for solutes considered in ternary systems

Solute	N	T-range (K)	P-range (bar)	r range	AARD (%)
anthracene	10	308-318	104-242	1.19-1.92	27.0
phenanthrene	10	308-318	104-242	1.19-1.92	15.5
2,6-dimethyl naphtalene	12	308-318	90-247	0.72-1.92	18.1
2,7-dimethyl naphtalene	12	308-318	90-247	0.72-1.92	16.9
1,10-decanediol	10	308-318	164-307	1.64-1.99	26.2

Comparison between the considered literature correlations and the proposed model was justified using comparative criteria. The model's performance is similar and sometimes superior to the literature models considered. The advantages of this model include the following: it does not require the knowledge of critical properties of the solutes and does take into account the binary interaction between solid solute and solvent. Moreover the predictive capabilities of the proposed model were well demonstrated both for solid-solvent and mixed solids-solvent systems.

REFERENCES

1.   G. Madras, C. Kulkarni and J. Modak, "Modeling the Solubilities of Fatty Acids in Supercritical Carbon Dioxide," Fluid Phase Equilibria, Vol. 209, No. 2, 2003, pp. 207-213. http://dx.doi.org/10.1016/S0378-3812(03)00148-1

2.   N. R. Foster, G. S. Gurdial, J. S. L. Yun, K. K. Liong, K. D. Tilly, S. S. T. Ting, H. Singh and J. H. Lee, "Significance of the Crossover Pressure in Solid Supercritical Fluid Phase-Equilibria," Industrial and Engineering Chemistry Research, Vol. 30, No. 8, 1991, pp. 1955-1964. http://dx.doi.org/10.1021/ie00056a044

3.   J. Chrastil, "Solubility of Solids and Liquids in Supercritical Gases," Journal of Physical Chemistry, Vol. 86, No. 15, 1982, pp. 3016-3021.http://dx.doi.org/10.1021/j100212a041

4.   D. L. Sparks, R. Hernandez and L. A. Estévez, "Evaluation of Density-Based Models for the Solubility of Solids in Supercritical Carbon Dioxide and Formulation of a New Model," Chemical Engineering Science, Vol. 63, No. 17, 2008, pp. 4292-4301. http://dx.doi.org/10.1016/j.ces.2008.05.031

5.   L. Nasri, Z. Bensetiti and S. Bensaad, "Correlation of the Solubility of Some Organic Aromatic Pollutants in Supercritical Carbon Dioxide Based on the UNIQUAC Equation," Energy Procedia, Vol. 18, 2012, pp. 1261-1270.http://dx.doi.org/10.1016/j.egypro.2012.05.142

6.   J. W. Lee, J. M. Min and H. K. Bae, "Solubility Measurement of Disperse Dyes in Supercritical Carbon Dioxide," Journal of Chemical and Engineering Data, Vol. 44, No. 4, 1999, pp. 684-687. http://dx.doi.org/10.1021/je9802930

7.   J. W. Lee, M. W. Park and H. K. Bae, "Measurement and Correlation of Dye Solubility in Supercritical Carbon Dioxide," Fluid Phase Equilibria, Vol. 173, No. 2, 2000, pp. 277-284. http://dx.doi.org/10.1016/S0378-3812(00)00404-0

8.   J. M. Prausnitz, R. N. Lichtenthaler and G. de Azevedo, "Molecular Thermodynamics of Fluid-Phase Equilibria," 3rd Edition, Prentice Hall Inc., Engelwood Cliffs, 1999.

9.   M. Vázquez da Silva and D. Barbosa, "Prediction of the Solubility of Aromatic Components of Wine in Carbon Dioxide," Journal of

Supercritical Fluids, Vol. 31, No. 1, 2004, pp. 9-25. http://dx.doi.org/10.1016/j.supflu.2003.09.018

10. R. Span and W. Wagner, "A New Equation of State for Carbon Dioxide Covering the Fluid Region from the Triple-Point Temperature to 1100 K at Pressures up to 800 MPa," Journal of Physical and Chemical Reference Data, Vol. 25, No. 6, 1996, pp. 1509-1596. http://dx.doi.org/10.1063/1.555991

11. S. E. Guigard and W. H. Stiver, "A Density-Dependant Solute Parameter for Correlating Solubilities in Supercritical Fluids," Industrial and Engineering Chemistry Research, Vol. 37, No. 9, 1998, pp. 3786-3792. http://dx.doi.org/10.1021/ie9702946

12. NIST database. http://webbook.nist.gov/chemistry/name-ser.html

13. K.-W. Cheng, M. Tang and Y.-P. Chen, "Calculations of Solid Solubility in Supercritical Fluids Using a Simplified Cluster Solvation Model," Fluid Phase Equilibria, Vol. 214, No. 2, 2003, pp. 169-186. http://dx.doi.org/10.1016/S0378-3812(03)00350-9

14. C. L. Laws, "Thermophysical Properties of Chemicals and Hydrocarbons," Wiliam Andrew Inc., 2008.

15. D. R. Lide, "CRC Handbook of Chemistry and Physics," 84th Edition, 2003-2004.

16. L. Li, Z.-C. Tan, S.-H. Meng and Y.-J. Song, "A Thermochemical Study of 1,10-Decanediol," Thermochim Acta, Vol. 342, No. 1-2, 1999, pp. 53-57. http://dx.doi.org/10.1016/S0040-6031(99)00305-6

17. M. Mukhopaday and G. V. Raghuram Rao, "Thermodynamique Modeling for Supercritical Fluid Process Design," Industrial & Engineering Chemistry Research, Vol. 32, No. 5, 1993, pp. 922-930. http://dx.doi.org/10.1021/ie00017a021

18. W. H. Stiver, S. E. Guigard and Beausoleil, "Predicting Ternary Solubilities Using a Solubility Parameter Approach," Proceedings of the 5th International Symposium on Supercritical Fluids, Atlanta, April 8-12, 2000.

19. S. Garnier, E. Neau, P. Alessi, A. Cortesi and I. Kikic, "Modelling Solubility of Solids in Supercritical Fluids Using Fusion Properties," Fluid Phase Equilibria, Vol. 158-160, 1999, pp. 491-500. http://dx.doi.org/10.1016/S0378-3812(99)00151-X

20. C.-C. Huang, M. Tang, W.-H. Tao and Y.-P. Chen, "Calculation of the Solid Solubilities in Supercritical Carbon Dioxide Using

a Modified Mixing Model," Fluid Phase Equilibria, Vol. 179, No. 1-2, 2001, pp. 67-84. http://dx.doi.org/10.1016/S0378-3812(00)00483-0

21. P. Coutsikos, K. Magoulas and D. Tassios, "Solubilities of Phenols in Supercritical Carbon Dioxide," Journal of Chemical and Engineering Data, Vol. 40, No. 4, 1995, pp. 953-958.http://dx.doi.org/10.1021/je00020a049

22. X. Wang and L. Tavlarides, "Solubility of Solutes in Compressed Gases: Dilute Solution Theory," Industrial and Engineering Chemistry Research, Vol. 33, No. 3, 1994, pp. 724-729. http://dx.doi.org/10.1021/ie00027a035

23. J. Méndez-Santiago and A. Teja, "The Solubility of Solids in Supercritical Fluids," Fluid Phase Equilibria, Vol. 158-160, 1999, pp. 501-510. http://dx.doi.org/10.1016/S0378-3812(99)00154-5

24. W. J. Schmitt and R. C. Reid, "Solubitity of Monofunctional Organic Solids in Chemically Diverse Supercritical Fluids," Journal of Chemical and Engineering Data, Vol. 31, No. 2, 1986, pp. 204-212. http://dx.doi.org/10.1021/je00044a021

25. A. Laitinen and M. Jaentti, "Solubility of 6-Caprolactam in Supercritical Carbon Dioxide," Journal of Chemical and Engineering Data, Vol. 41, No. 6, 1996, pp. 1418- 1420.http://dx.doi.org/10.1021/je9600313

26. K. J. Pennisi and E. H. Chimowitz, "Solubilities of Solid 1,10-Decanediol and a Solid Mixture of 1,10-Decanediol and Benzoic Acid in Supercritical Carbon Dioxide," Journal of Chemical and Engineering Data, Vol. 31, No. 3, 1986, pp. 285-288.http://dx.doi.org/10.1021/je00045a008

27. Z. Eckert, "Correlation and Prediction of Solid-Supercritical Fluid Phase Equilibria," Industrial & Engineering Chemistry Process Design and Development, Vol. 22, No. 4, 1983, pp. 582-588. http://dx.doi.org/10.1021/i200023a005

28. Z. Huang, S. Kawi and Y. C. Chiew, "Application of the Perturbed Lennard-Jones Chain Equation of State to Solute Solubility in Supercritical Carbon Dioxide," Fluid Phase Equilibria, Vol. 216, No. 1, 2004, pp. 111-122.http://dx.doi.org/10.1016/j.fluid.2003.10.004

29. A. Delle Site, "The Vapor Pressure of Environmentally Significant Organic Chemicals: A Review of Methods and Data at Ambient

Temperature," Journal of Physical and Chemical Reference Data, Vol. 26, No. 1, 1997, p. 157. http://dx.doi.org/10.1063/1.556006

30. V. Oja and E. M. Suuberg, "Vapor Pressures and Enthalpies of Sublimation of Polycyclic Aromatic Hydrocarbons and Their Derivatives," Journal of Chemical and Engineering Data, Vol. 43, No. 3, 1998, pp. 486-492. http://dx.doi.org/10.1021/je970222l

31. K. Tochigi, T. Iizumi and K. Kojima, "High Pressure Vapor-Liquid and Solid-Gas Equilibria," Industrial and Engineering Chemistry Research, Vol. 37, No. 9, 1998, pp. 3731-3740. http://dx.doi.org/10.1021/ie970060m

32. E. Kosal and G. D. Holder, "Solubility of Anthracene and Phenanthrene Mixtures in Supercritical Carbon Dioxide," Journal of Chemical and Engineering Data, Vol. 32, No. 2, 1987, pp. 148-150. http://dx.doi.org/10.1021/je00048a005

33. Y. Iwai, Y. Mori, N. Hosotani, H. Higashi, T. Furuya, Y. Arai, K. Yamamoto and Y. A Mito, "Solubilities of 2,6-and 2,7-Dimethylnaphthalenes in Supercritical Carbon Dioxide," Journal of Chemical and Engineering Data, Vol. 38, No. 4, 1993, pp. 509-511.http://dx.doi.org/10.1021/je00012a006

34. J. F. Brenneck and C. A. Eckert, "Phase Equilibria for Supercritical Fluid Process Design," AIChE Journal, Vol. 35, No. 9, 1989, pp.1409-1427.http://dx.doi.org/10.1002/aic.690350902

35. R. T. Kurnik and R. C. Reid, "Solubility of Solid Mixtures in Supercritical Fluids," Fluid Phase Equilibria, Vol. 8, No. 1, 1982, pp. 93-105. http://dx.doi.org/10.1016/0378-3812(82)80008-3

36. J. Kwiatkowski, Z. Lisicki and W. Majewski, "An Experimental Method for Measuring Solubilities of Solids in Supercritical Fluids," Berichte der Bunsengesellschaft für physikalische Chemie, Vol. 88, No. 9, 1984, pp. 865-869.http://dx.doi.org/10.1002/bbpc.19840880919

Antioxidant Properties and Chemical Composition Relationship of Europeans and Brazilians Propolis

Sabrina Fabris[1], Mariangela Bertelle[1], Oxana Astafyeva[2], Elena Gregoris[3], Roberta Zangrando[3], Andrea Gambaro[4], Giuseppina Pace Pereira Lima[5], and Roberto Stevanato[1]

[1]Department of Molecular Sciences and Nanosystems, University Ca' Foscari of Venice, Venice, Italy;

[2]Depatment of Biotechnology and Bioecology, Astrakhan State University (ASU), Astrakan, Russia;

[3]Institute for the Dynamics of Environmental Processes, CNR, University Ca' Foscari of Venice, Venice, Italy;

[4]Department of Environmental Sciences, Informatics and Statistics, University Ca' Foscari of Venice, Venice, Italy;

[5]Institute of Biosciences, São Paulo State University, Botucatu, Brazil

ABSTRACT

The antioxidant activity of ethanol extracts of propolis, bee glue, of various climate and orographic characteristics, collected from Italy, Brazil and Russia, was evaluated measuring their inhibitory action on peroxidation of linoleic acid, radical scavenging ability towards 2,2'-diphenyl-1-picrylhydrazyl, total phenolic content and reducing capacity by enzymatic and Folin method respectively. Propolis samples were chemically characterized by HPLC-MS/MS in order to find a possible correlation between antioxidant activity and polyphenols composition and quantification. The results obtained indicate that Italian and Russian propolis samples have similar polyphenolic composition and, as a consequence, almost similar antioxidant activity, while Brazilian propolis evidence lower polyphenolic and antioxidant characteristics. Climate and orography reasons of these differences are also suggested.

INTRODUCTION

Propolis (bee glue, CAS No. 9009-62-5) is a resinous natural product collected by honeybees from various plant sources, characterized by antiseptic, antimycotic, bacteriostatic, astringent, spasmolytic, anti-inflammatory, anaesthetic and antioxidant properties [1-3].

The chemical composition of propolis (polyphenols, terpenoids, steroids and amino acids contents) can be very different, depending on the place of collection [4,5] and, in particular, on the composition of the plant source.

In continental Europe, Populus nigra, the black poplar, is the source of choice for bees; in Russia, especially in northern parts, is Betula verrucosa [6], while in South America is Hyptis divaricata and Baccharis dracunculifolia [7].

Particularly interesting is propolis from Brazil for the vast biodiversity of the country [8]; within the Brazilian territory, which includes temperate, subtropical and tropical zones, 12 different types of propolis have been classified, depending on their composition and botanical origin: five from the south, six from the northeast and one from the south-east named propolis "green" [7].

As a consequence, propolis samples from Europe and Brazil have different chemical composition [9]; in particular, flavonoids and phenolic acid esters are the main components of European propolis, representing about 10% - 15% of the weight [10,11], while in Brazilian propolis the major components are terpenoids and prenylated derivatives of p-coumaric acids, being flavonoids and phenolic acid esters amount < 4%, [1,12,13].

Despite the different composition, the two types of propolis have very similar pharmacological properties, so that some studies suggested that other compounds other than flavonoids, as aromatic acids contained in high amount, could be responsible for the antifungal effects [8].

On the contrary, few studies report the antioxidant activity of Brazilian propolis and only one [14] compares propolis originating from temperate and tropical zones.

It appears therefore attractive to verify if a reduced presence of flavonoids matches a lower antioxidant activity or whether, as it happens with other properties, different compounds become substitutes for flavonoids in their action against free radicals.

The purpose of this study is to compare the antioxidant activity of propolis from Europe and Brazil, to relate it to their chemical composition and to individuate the species responsible of antioxidant capacity.

From European propolis we focalized our attention on some Italian and Russian samples. These last are of particular interest because picked in sites with a vast biodiversity and because their composition and antioxidant activity are unknown in literature.

We consider only one kind of Italian propolis, because as seen in a previous work [11] they are quite similar for chemical composition, UV spectra and antioxidant activity.

For the lack of a universal and unique method to determine the antioxidant activity of a compound, we studied the inhibitory action of propolis on lipid peroxidation of linoleic acid (LA) which, in our opinion, mimes better than other methods the efficacy of an antioxidant compound to prevent oxidative damages on lipoproteins or cell membrane by ROS injures. Then the results were compared with the radical scavenging ability towards 2, 2′-diphenyl-1-picrylhydrazyl (DPPH method), the reducing capacity by the Folin Ciocalteu assay and the total phenolic content using the enzymatic method [15].

EXPERIMENTAL

Material

Raw native propolis was obtained directly from beekeepers and conserved in closed vessels at 3°C to prevent natural oxidation.

Italian propolis comes from Montello hill hood, located in the Veneto region, at the boundary between Padana plain and Alpi mountains. The three Russian propolis comes from Sochi, a city situated on the shores of the Black Sea, near the Caucasus mountains; Volgograd, a city of the European Russia, along the Volga River; Dagestan, a Republic located in the North Caucasus mountains and bordered on its eastern side by the Caspian Sea.

Brazilian propolis are collected in Pantanal, a tropical wetland in the state of Mato Grosso; Botucatu, a city in the southeastern region of Brazil located in the State of São Paulo; "green" propolis was from Minas Gerais, a state in the west of the southeastern subdivision of Brazil.

Köppen climate classification and the orography characteristics of the origin zones are reported in Table 1.

All experiments were repeated at least in triplicate and carried out in different times.

Chemicals

All chemicals were analytical grade and were supplied from Sigma Chemical Co. (USA) or Romil Ltd. for HPLC-MS/MS measurements. ABIP (2, 2'-azobis [2'-(2- imodazolin-2-yl) propane] dihydrochloride) was obtained from Wako Chemicals (Germany). The aqueous solutions were prepared with quality milliQ water.

Preparation of Ethanolic Extract of Propolis

Ethanolic extract of propolis (EEP) was obtained dissolving raw propolis samples overnight under vigorous agitation at 3°C. After filtration through a strainer to remove insoluble residual beehive products, i.e. wood fragments, bee bodies, etc., the suspension was left to sediment and the supernatant was centrifuged for 30 min at 2000 rpm. Limpid solution, without further purifications, was used for successive analyses. Solution concentration was calculated weighting dry residue after complete evaporation of all solvent until dryness.

Table 1: Köppen climate classification and Orography characteristics of the origin zones

Sample	Origin zone	Köppen climate classification	Orography characteristics of the origin zones
MONT	Montello, Veneto, Italy	Mediterranean-marine west coast boundary	Hill hood
DAGE	Dagestan, Caucasus, Russia	moderate continental-arid	Plains-mountains
VOLG	Volgograd, Russia	semi-arid (eastern part) or forest-steppe (north-western part)	Plains
SOCHI	Sochi, Krasnodar, Russia	humid subtropical	Sea-hill
MG	Minas Geraris, Brazil	Tropical-tropical of altitude	hill
SP1	Botucatu, São Paulo Brazil	subtropical	hill
SP2	Botucatu, São Paulo Brazil	subtropical	hill
MT1	Pantanal, Mato Grosso, Brazil	humid tropical	tropical wetland
MT2	Pantanal, Mato Grosso, Brazil	humid tropical	tropical wetland

UV Measurements

Spectrophotometric measurements were recorded on a UV-VIS Shimadzu UV-1800 instrument equipped with a temperature controlled quartz cell. Specific absorbance ($E_{1cm}^{1\%}$) of each EEP sample was obtained according to the method of Miyataka [16].

Inhibition of Lipid Peroxidation (ILP)

The antioxidant activity of propolis to prevent linoleic acid (LA) peroxidation was studied in sodium dodecyl sulfate (SDS) micelles. As previously reported [17], the propolis antioxidant capacity was calculated as the antioxidant concentration (mg/L) that halves the rate of oxygen consumption due to the peroxidation process and it is expressed as inhibitory concentration (IC_{50}).

DPPH (2, 2-Diphenyl-1-picrylhydrazyl) Radical Scavenging Capacity Assay

This method is based on the capacity of an antioxidant to scavenge the stable free radical DPPH [18]. The procedure is reported in Stevanato [15] and the results are expressed as catechin equivalent (CE).

Folin Ciocalteu Assay and Total Phenolics Content (TPC)

The Folin Ciocalteu assay and the Total Phenolic Content determination by enzymatic method were carried out spectrophotometrically, according to the procedures previously described [15] and the results are expressed as catechin equivalent (CE).

Electrochemical Measurements

The measures of oxygen consumption were performed by a potentiostat Amel 559, equipped with an oxygen microelectrode Microelectrodes MI-730.

HPLC-MS/MS Measurements

High performance liquid chromatography with triple quadrupole mass spectrometry detection (HPLC/ (-) ESI-MS/MS) was used to identify the constituents of propolis. HPLC analyses of ethanolic extracts of propolis (EEP), diluted with methanol and filtered with a 0.45 μm filter, were carried out by an Agilent 1100 HPLC system (Agilent Technologies, USA). For the chromatographic analysis, 5 μl of the sample were injected onto a C18 Synergy Hydro-RP 80A column (50 × 2 mm, 4 μm particle size) using an Aqua C18 125A pre-column (2 mm i.d. × 4 mm length). The mobile phase was acetic acid 0.1% (A) and MeOH (B). The gradient was: 10% - 90% B (2 min), 90% - 97% B (2 - 9 min), 97% - 100% B (9 - 10 min), 100% B (10 - 15 min) at a flow rate of 250 μl/min An API 4000 triple quadrupole mass spectrometer (Applied Biosystems /MDS SCIEX, Toronto, Canada) equipped with a Turbo V™ source was used to detect polyphenols in propolis. Calibration curves of peak area versus analyte concentration were plotted for the studied polyphenols using the standard addition technique [19]. All data were acquired in negative ionization mode by multiple reactions monitoring (MRM).

RESULTS AND DISCUSSION

UV Spectra and Specific Absorbance

Figure 1 shows the absorbance spectra of equal concentrations of different EEP samples; in Table 2, the specific absorbances ($E_{1cm}^{1\%}$) at the wavelength of absorption maximum (λ_{max}) are also reported.

As previously reported, Italian propolis shows a strong absorption in the region between 250 and 400 nm, with a very intense peak at 290 nm and a shoulder between 320 and 330 nm. This profile is compatible with that of the flavonoids, which generally show a first maximum between 240 and 285 nm, due to absorption of the ring A, and another maximum, of variable position, above 300 nm, depending on the substitution and conjugation of ring C [20]. The spectrum is characterized by a very high specific absorbance, higher than that of propolis from other countries, reported here and above [14].

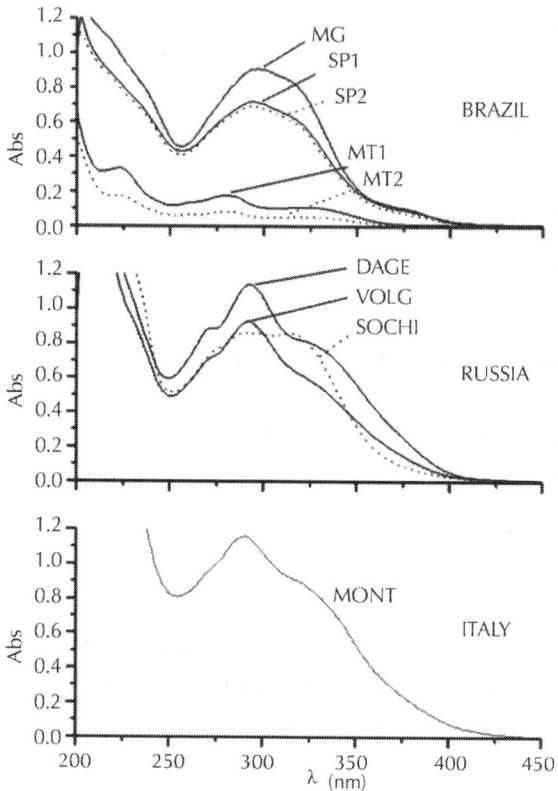

Figure 1: Absorption spectra of EEP samples. Concentration of all samples was 30 mg/L.

Table 2: Specific absorbance $(E_{1cm}^{1\%})$ and wavelength of absorption maximum (λ_{max}) of EEP samples

Sample	Specific absorbance $(E_{1cm}^{1\%})$	Wavelength of absorption maximum (λ_{max}) (nm)
MONT	390	290
DAGE	380	292
VOLG	309	292
SOCHI	286	299
	286	314

MG	302	295
SP1	238	295
SP2	230	295
MT1	111	223
	59	280
MT2	57	223
	27	280

The UV spectra of Russian propolis are quite similar to Italian one, excluding SOCHI sample which shows two peaks with the same intensity at 299 and 314 nm, apparently similar, for profile and intensity, to the spectra of MG and SP Brazilian samples. In particular, DAGE sample has a specific absorbance close to Italian one, while VOLG appears smaller.

The UV spectra of Brazilian propolis are different: all show an absorption of ultraviolet radiation between 250 and 400 nm, but the samples MG, SP1 and SP2 have at all the wavelengths absorbance values substantially high, with a well-defined peak at 295 nm and a shoulder at 315 nm; MT1 and MT2 samples, instead, have a different profile of lower intensity, with a peak around 280 nm and a shoulder at 325 nm. In the first analysis, we can therefore hypothesize that among the five samples, two different types of propolis can be individualized.

By a general comparison, it emerges that MONT and DAGE samples show similar profile and specific absorbance, while VOLG appears of lower absorbance intensity; MG and SP are comparable, but the first has higher specific absorbance; SOCHI has specific absorbance comparable to VOLG but shows two peaks of equal intensity instead of a peak and a shoulder; MT samples show different profiles and a much lower molar absorbance (about 10% of the Italian one). These lower absorbance values recorded in the range 250 - 400 nm indicate a significant lower content of flavonoids, taking into account that flavonoids absorb in this wavelength range.

Antioxidant Capacity

The graph of Figure 2, where the lipid peroxidation inhibition property of propolis samples, expressed as IC_{50}, is reported, evidences completely different behavior between European and Brazilian propolis. In fact,

Italian and Russian propolis show similar values characterized by high antioxidant capacity (IC_{50} < 1 mg/L) if compared to the Brazilian ones, which IC_{50} values (from about 2 to 20 mg/L) range between 4 to 40 times higher than the average of European samples.

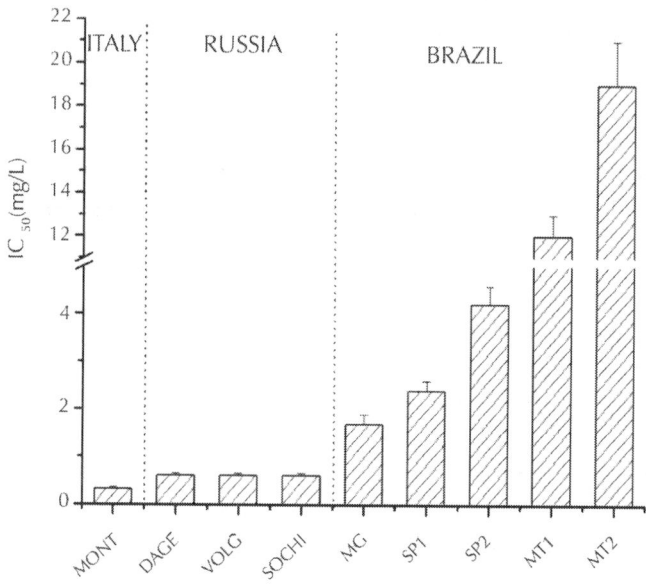

Figure 2: IC_{50} (50% inhibitory concentration) values of lipid peroxidation inhibition of examined EEP samples.

In particular, while IC_{50} values of Italian and Russian propolis are comparable (0.3 - 0.6 mg/L), Brazilian samples display high variability: propolis from the Pantanal (MT1 and MT2) differs from the other because of its high IC_{50} (12 - 19 mg/L), followed by samples from São Paulo characterized by an intermediate IC_{50} values (2.4 - 4.2 mg/L). Among these, green propolis (MG) seems to possess a higher antioxidant activity (1.7 mg/L), but not so different from SP samples and, in any case, about four times lower than European propolis.

These results are in agreement with specific absorbance data, according to considerations previously reported [11]: a propolis of greater specific absorbance matches a greater antioxidant activity; as

a consequence, $(E_{1cm}^{1\%})$ can be considered a preliminary method to test the antioxidant activity of EEP.

From Figure 3, where the results of Folin, enzymatic and DPPH assays are reported, it appears that data obtained with DPPH and enzymatic method are comparable and in agreement with lipid peroxidation results and substantiate the different antioxidant activity between European and Brazilian propolis. On the contrary, no significant difference between European and Brazilian samples is observed using Folin assay, confirming the a-specificity of this method [21].

HPLC/MS Measurements

The concentration of main polyphenolic components of EEP samples obtained with HPLC/MS spectrometry is reported in Figure 4. The graph shows a marked difference between the polyphenols concentration of European and Brazilian propolis: the average concentration of polyphenols of Italian and Russian propolis is comparable (with the exception of SOCHI sample), while that of the Brazilian one is at least one magnitude order times lower (see the different y-axis scale).

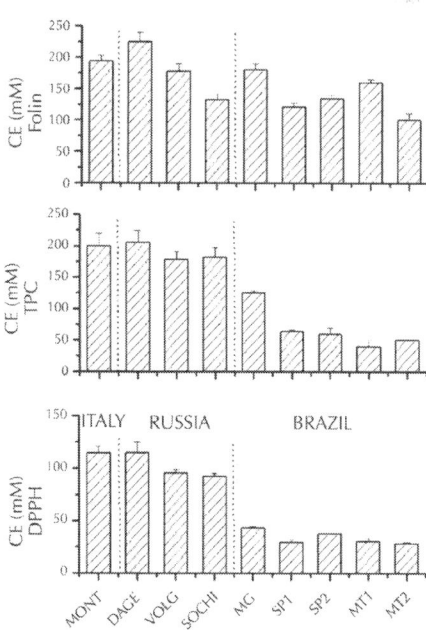

Figure 3: CE (catechin equivalent) values of examined EEP samples obtained by DPPH, enzymatic (TPC) and Folin assays.

In the Italian propolis, chrysin and pinocembrin are present in high concentration (≥40 mg/g propolis), followed by galangin, CAPE (caffeic acid phenethyl ester), caffeic acid, and DMAC (1, 1-dimethylallylcaffeate), all ranging from 7 to 18 mg/g propolis approximately, and with very low amounts, from 1 to 3 mg/g propolis, of remaining polyphenols.

In the Russian propolis, DAGE sample shows highest polyphenols concentration, followed by VOLG, while SOCHI sample shows concentrations at least a magnitude order lower. In DAGE sample, galangin and pinocembrin concentrations appear similar (»60 mg/g propolis), such as DMAC and chrysin (»40 mg/g propolis), while caffeic acid concentration is similar to that of Montello sample.

Volgrad and Montello samples show comparable chemical composition. As previously found [11], chrysin and pinocembrin are characterized by low antioxidant activeity, so the antioxidant activity of European propolis is due above all to galangin, CAPE, caffeic acid and DMAC.

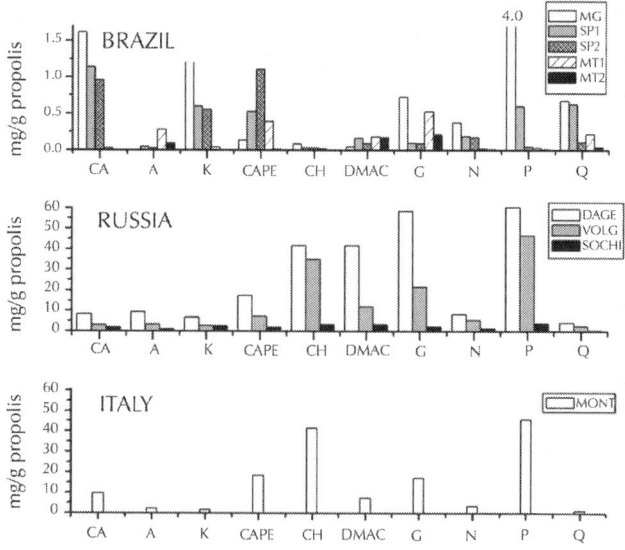

Figure 4: Concentration of main polyphenolic compounds of EEP samples obtained with HPLC/MS. CA: caffeic acid, A: apigenin, K: kaempferol, CAPE: caffeic acid phenylethyl ester, CH: crysin, DMAC: 1, 1-dimethylallylcaffeate, G: galangin, N: naringenin, P: pinocembrin, Q: quercetin.

On the basis of their chemical composition, the Brazilian propolis can be divided ideally into three classes which correspond, as expected, to their origin zone.

The propolis containing the largest amount of total polyphenols comes from Minas Gerais, followed by São Paulo's samples, while Mato Grosso samples show very low total polyphenols concentrations.

With regard to the individual polyphenols found, singularly Minas Gerais sample shows a peak of pinocembrin (»4 mg/g propolis) and significant amounts of caffeic acid, kaempferol, galangin and quercetin (0.7 - 1.6 mg/g propolis); the São Paulo's samples contain mainly caffeic acid (»1 mg/g propolis), kaempferol and CAPE (»0.6 mg/g propolis), with, in the case of the SP1 sample, also non-negligible amount of pinocembrin and quercetin (»0.6 mg/g propolis) have been found. The Mato Grosso samples contain mainly apigenin, galangin and DMAC (0.2 - 0.5 mg/g propolis) and in the case of MT1, a significant amount of CAPE and quercetin (0.2 - 0.5 mg/g propolis).

These samples differ from each other not only for the total amount of polyphenols content, but also for the relationships between the individual related polyphenols, contrary to what found for the Italian propolis [11]. These differences are linked to the very high biodiversity of Brazilian territory, where different types of plants containing different chemical compounds are present.

With regard to the European Propolis, conversely, since the chemical composition of Dagestan and Volgograd is comparable to that of Montello, we can assume that in these Russian area the main plant source is Populus nigra and not Betula verrucosa as reported for North Russia, while a different source can be hypothesized for propolis from Sochi, that shows chemical composition and UV-VIS spectrum more similar to that of São Paulo. This can be explained with the subtropical climate of Sochi, more similar to that of São Paulo rather than to that other part of Russia. As a consequence, the plant sources of Sochi's propolis could be more similar to that of São Paulo. However Sochi sample shows antioxidant activity higher than that of all Brazilian propolis and comparable to that of other European propolis; in this case, perhaps, other compounds, different from analyzed polyphenols, could be responsible of its antioxidant activity.

In the case of other samples, conversely, there is a good agreement between concentrations of the various polyphenols found with HPLC-

MS/MS analysis and results of antioxidant property and specific absorbance: to a higher content of polyphenols correspond a higher absorbance value and a higher antioxidant power. This means that for all samples, with the only exception of that from Sochi, the antioxidant activity is due to analyzed polyphenols. At last, Sochi sample needs more thorough investigation in order to clarify the chemical composition reasons of its high antioxidant activity.

REFERENCES

1.   M. C. Marcucci and V. Bankova, "Chemical Composition, Plant Origin and Biological Activity of Brazilian Propolis," Current Topics in Phytochemistry, Vol. 2, 1999, pp. 115-123.

2.   G. A. Burdock, "Review of the Biological Properties and Toxicity of Bee Propolis," Food and Chemical Toxicology, Vol. 36, No. 4, 1998, pp. 347-363. doi:10.1016/S0278-6915(97)00145-2

3.   A. H. Banskota, Y. Tezuka and S. Kadota, "Recent Progress in Pharmacological Research of Propolis," Phytotherapy Research, Vol. 15, No. 7, 2001, pp. 561-571.doi:10.1002/ptr.1029

4.   E. L. Ghisalberti, P. R. Jefferies, R. Lanteri and J. Matisons, "Constituents of Propolis," Experientia, Vol. 34, No. 2, 1978, pp. 157-158. doi:10.1007/BF01944648

5.   V. Bankova, A. Dyulgerov, S. Popov, L. Evstatieva, L. Kuleva, O. Pureb and Z. Zamjansan, "Propolis Produced in Bulgaria and Mongolia: Phenolic Compounds and Plant Origin," Apidologie, Vol. 23, No. 1, 1992, pp. 79-85. doi:10.1051/apido:19920109

6.   S. A. Popravko and M. V. Sokolov, "Plant Sources of Propolis," Pchelovodstvo, Vol. 2, 1980, pp. 28-29.

7.   Y. K. Park, S. M. Alencar, C. L. Aguiar, "Botanical Origin and Chemical Composition of Brazilian Propolis," Journal of Agricolture and Food Chemistry, Vol. 50, No. 9, 2002, pp. 2502-2506. doi:10.1021/jf011432b

8.   K. Salomão, A. P. Dantas, C. M. Borba, L. C. Campos, D. G. Machado, F. R. Aquino Neto and S. L. De Castro, "Chemical Composition and Microbicidal Activity of Extracts from Brazilian and Bulgarian Propolis," Letters of Applied Microbiology, Vol. 38, No. 2, 2004, pp. 87-92. doi:10.1111/j.1472-765X.2003.01458.x

9. M. C. Marcucci, "Propolis: Chemical Composition, Biological Properties and Therapeutic Activity," Apidologie, Vol. 26, No. 2, 1995, pp. 83-99. doi:10.1051/apido:19950202

10. V. S. Bankova, S. L. D. Castro and M. C. Marcucci, "Propolis: Recent Advances in Chemistry and Plant Origin," Apidologie, Vol. 31, No. 1, 2000, pp. 3-15.doi:10.1051/apido:2000102

11. E. Gregoris and R. Stevanato, "Correlations between Polyphenolic Composition and Antioxidant Activity of Venetian Propolis," Food Chemical Toxicology, Vol. 48, No. 1, 2010, pp. 76-82. doi:10.1016/j.fct.2009.09.018

12. S. Tazawa, T. Warashina, T. Noro and T. Miyase, "Studies on the Constituents of Brazilian Propolis," Chemical & Pharmaceutical Bullettin, Vol. 46, 1998, pp. 1477- 1479.doi:10.1248/cpb.46.1477

13. S. Tazawa, T. Warashina and T. Noro, "Studies on the Constituents of Brazilian Propolis II," Chemical & Pharmaceutical Bullettin, Vol. 47, No. 10, 1999, pp. 1388-1392.

14. S. Kumazawa, T. Hamasaka and T. Nakayama, "Antioxidant Activity of Propolis of Various Geographic Origins," Food Chemistry, Vol. 84, No. 3, 2004, pp. 329- 339.doi:10.1016/S0308-8146(03)00216-4

15. R. Stevanto, S. Fabris and F. Momo, "New Enzymatic Method for the Determination of Total Phenolic Content in Tea and Wine," Journal of Agricolture and Food Chemistry, Vol. 52, No. 20, 2004, pp. 6287-6293. doi:10.1021/jf049898s

16. H. Miyataka, M. Nishiki, H. Matsumoto, T. Fujimoto, M. Matsuka and T. Satoh, "Evaluation of Propolis. I. Evaluation of Brazilian and Chinese Propolis by Enzymatic and Physico-Chemical Methods," Biological & Pharmaceutical Bullettin, Vol. 20, No. 5, 1997, pp. 496-501.

17. S. Fabris, F. Momo, G. Ravagnan and R. Stevanato, "Antioxidant Properties of Resveratrol and Piceid on Lipid Peroxidation in Micelles and Monolamellar Liposomes," Biophysical Chemistry, Vol. 135, No. 1, 2008, pp. 76-83. doi:10.1016/j.bpc.2008.03.005

18. M. S. Blois, "Antioxidant Determinations by the Use of a Stable Free Radical," Nature, Vol. 181, No. 4617, 1958, pp. 1199-1200. doi:10.1038/1811199a0

19. K. Danzer and L. A. Currie, "Guidelines for Calibration in Analytical Chemistry," Pure & Applied Chemistry, Vol. 70, No. 4, 1998, pp. 993-1014. doi:10.1351/pac199870040993

20. E. De Rijke, P. Out, W. M. A. Niessen, F. Ariese, C. Gooijer and U. A. T. Brinkman, "Analytical Separation and Detection Methods for Flavonoids," Journal of Chromatography A, Vol. 1112, No. 1-2, 2006, pp. 31-63. doi:10.1016/j.chroma.2006.01.019

21. R. Stevanato, S. Fabris, M. Bertelle, E. Gregoris and F. Momo, "Phenolic Content and Antioxidant Properties of Fermenting Musts and Fruit and Vegetable Fresh Juices," Acta Alimentaria, Vol. 38, No. 2, 2009, pp. 193-203. doi:10.1556/AAlim.2008.0031

Influence of Bath Temperature, Deposition Time and [S]/[Cd] Ratio on the Structure, Surface Morphology, Chemical Composition and Optical Properties of CdS Thin Films Elaborated by Chemical Bath Deposition

Fouad Ouachtari[1], Ahmed Rmili[1], Sidi El Bachir Elidrissi[1], Ahmed Bouaoud[1], Hassan Erguig[1], and Philippe Elies[2]

[1]Physics Laboratory Matter and Radiation, Team Optical Spectroscopy of Solid Matter, Physics Department, Faculty of Sciences, Kenitra, Morocco

[2]Platform for Imaging and Measurement in Microscopy 6, av. Brest Cedex, France

ABSTRACT

Cadmium sulphide (CdS) thin films were deposited on glass substrates by the chemical bath deposition (CBD) method, using anhydrous cadmium chloride (CdCl$_2$) and thiourea (CS(NH$_2$)$_2$) as sources of cadmium and sulphur ions respectively. The influence of bath temperature (T$_b$), deposition time (t$_d$) and [S]/[Cd] ratio in the solution on the structural, morphological, chemical composition and optical properties of these films were investigated. XRD studies revealed that all the deposited films were polycrystalline with hexagonal structure and exhibited (002) preferential orientation. The films deposited under optimum conditions (T$_b$ = 75°C, t$_d$ = 60 min and [S]/[Cd] ratio = 2.5) were relatively well crystallized. These films showed large final thickness and their surface morphologies were composed of small grains with an approximate size of 20 to 30 nm and grains grouped together to form large clusters. EDAX analysis revealed that these films were nonstoichiometric with a slight sulphur deficiency. These films exhibited also a transmittance value about 80% in the visible and infra-red range.

INTRODUCTION

Amongst of the chalcogenide thin films like PbS, CdS, ZnS and MnS, CdS appear as an interesting material for using as n-type window layer for p-CdTe and chalcopyritebased solar cells such as p-CuInSe$_2$, and/or p-Cu(In,Ga)Se$_2$ (CIGS) [1]. This is because CdS has high transparency, wide and direct band gap transition (2.42 eV), photoconductivity, high electron affinity and n-type conductivity. CdS can also be used in a lot of applications including electronic [2] and optoelectronic devices [3]. Undoped and doped CdS thin films have been reported using different methods: electrodeposition (ED) [4], spray pyrolysis (SP) [5], chemical bath deposition (CBD) [6], molecular beam epitaxy (MBE) [7], metal organic vapour phase epitaxy (MOVPE) [8], successive ionic layer adsorption and reaction (SILAR) [9], and physical vapour deposition (PVD) [10]. Among these methods, the CBD technique is relatively simple, low cost compared to other methods requiring vacuum environment and its capable to yield films with good quality at optimum growth conditions.

The aim of this present work is to study the influence of some deposition parameters, such as bath temperature, deposition time and [S]/[Cd] ratio in the solution (thiourea to cadmium chloride concentration) on the crystalline structure, surface morphology, chemical composition and optical properties of CdS thin films prepared by chemical bath deposition.

EXPERIMENTAL PROCEDURE

CBD is a technique in which thin films are deposited on substrates immersed in dilute alkaline solution containing metal ions and the chalcogenide source. This method of deposition usually uses a complexing agent to control the slow release of metal ions (Cd^{2+}) and sulphur ions (S^{2-}) to produce the controlled homogeneous precipitation of the film on the solid substrate. When the complexing agent is ammonia (NH_3), the possible chemical reactions to form CdS films are as follows [11]:

$$Cd(NH_3)_4^{2+} \leftrightarrow 4NH_3 + Cd^{2+} \tag{1}$$

$$CS(NH_2)_2 + OH^- \leftrightarrow SH^- + CH_2N_2 + H_2O \tag{2}$$

$$SH^- + OH^- \leftrightarrow S^{2-} + H_2O \tag{3}$$

$$Cd^{2+} + S^{2-} \leftrightarrow CdS \tag{4}$$

In this work, the initial solutions to elaborate CBDCdS films are prepared from anhydrous cadmium chloride ($CdCl_2$), thiourea ($CS(NH_2)_2$), ammonia (NH_3), ammonium chloride (NH_4Cl). Cadmium chloride of 0.12 M and thiourea of 0.3 M are employed as the cadmium and the sulfur sources, respectively. Ammonia of 10 M is used as a complexing agent. Firstly, 3.75 ml cadmium chloride solution is added to 112.5 ml of de-ionized water. Thereafter, 15 ml ammonia solution is added at the same time with 15 ml ammonium chloride solution of 0.01 to 2 M to adjust the pH at about 10 under the control of a pH meter. Pre-treated commercial microscope slides (1.5 × 2.5 cm^2) are inserted vertically into the bath and the solution is heated at appropriate temperature (between 60°C and 90°C). Finally, when a desired temperature is obtained, 3.75 ml of thiourea solution is added under stirring condition to ensure homogeneous distribution of the chemicals. The total volume of solution is 150 ml. After deposition, the substrates are removed from the chemical bath, and cleaned for several times with de-ionized water, then dried in air. The formed films are yellow in colour and exhibit good adherence to the substrate surfaces.

The crystal phase of the films is determined by X-ray diffraction (XRD), using CuK radiation with 2 ranging from 10° to 70°. The surface morphology of the films is determined by scanning electron microscopy (SEM) and atomic force microscopy (AFM). The chemical composition is performed with an EDAX spectrometer attached to the scanning electron microscope. In order to determine the band gap energy of the films, the optical transmission study is carried out in the wavelength range of 300 to 2500 nm, using a SHIMADZU 3101 PC UV-VISNIR spectrophotometer. The thicknesses are measured by the gravimetric method with accuracy of 10%.

RESULTS AND DISCUSSION

Crystal Structure Determination

Figure 1 shows the X-Ray diffraction diagrams of the CdS thin films deposited at bath temperature varying from 60 to 90°C and deposition time of 60 min. A single diffraction peak at 2q = 26.7° is observed. The interplanar spacing values corresponding to this diffraction peak (d$_{hkl}$ = 3.34 Å) is compared with the ASTM DATA [12]. This suggest that the obtained films are crystallized in the hexagonal structure with a preferred orientation along the (002) direction. However, R. Zhai et al. [13], O. Oladeji et al. [14] and M. Ichimura et al. [15], using chemical bath deposition and other techniques, have been found a cubic structure of the CdS films. In fact, depending on the preparation method, cadmium sulphide can exist in both sphalerite cubic and hexagonal forms, but the latter structure is more stable [6]. In addition the hexagonal structure of the CdS films is preferable to use in solar cell applications because the lattice parameter mismatch with $CuInSe_2$(1.2%) compared to that of cubic CdS (0.7%). Moreover, Figure 1 shows that the whole X-Ray diffraction diagrams exhibit a broad hump near the (002) peak at $2q$ = 26.7°, which is due to the glass substrate. The intensity of the (002) plane is found to be increased when increasing bath temperature up to 75°C, which decreased afterwards. This suggests that the crystallinity of these films increases when the bath temperature increases; which is explained by the increase in films thickness due to decomposition of reactants and production of ions which is necessary for films formation. The decrease of the (002) plane intensity for films deposited at bath temperature greater than 75°C indicates a deterioration of the crystallinity which is attributed to the relatively lower thicknesses of the films resulted (Figure 2). The decrease of the films thickness at this temperature range is explained by the dissolution of the preformed CdS films and the desorption phenomenon [16]. The lattice parameters a and c are calculated from the peaks positions using the formula of hexagonal system. The values are found to be a = 4.13 Å, c = 6.70 Å and c/a = 1.63 which are close to the values published in the literature [17].

$$\frac{1}{d_{hkl}^2} = \frac{4\left(h^2 + k^2 + hk\right)}{3a^2} + \frac{l^2}{c^2}$$

(5)

The average crystallite sizes (D_{hkl}) of the CdS films are estimated from the X-ray diffraction patterns using the Scherrer formula [18]: where   is the wavelength of incident radiation (λ = 1.544 Å), β_{hkl} is the full-width at half maximum (FWHM) of the respective diffraction peak and q_{hkl} is the Bragg diffraction angle.

The calculated values are reported in Table 1. As it can be seen the values are found in the nanometer region (15 - 21 nm), indicating that the polycrystalline CdS films are made up of nanocrystal particles.

$$D_{hkl} = 0.9 \frac{\lambda_{hkl}}{\beta_{hkl} \cos(\theta_{hkl})}$$

(6)

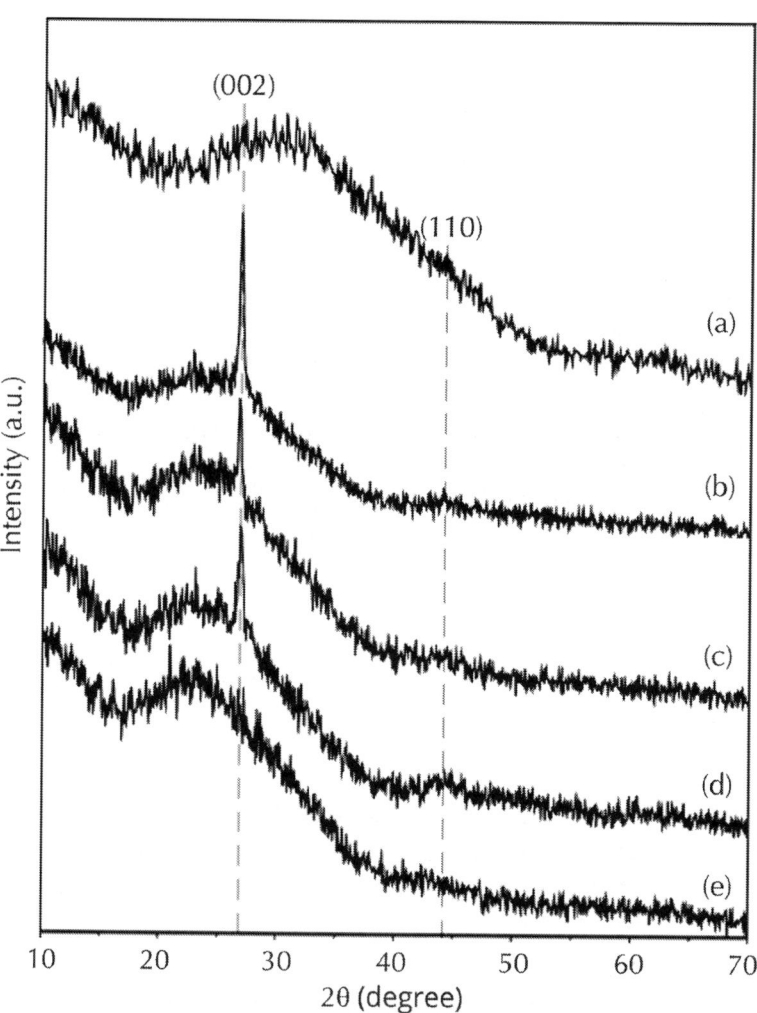

Figure 1: X-ray diffraction patterns of CdS thin films deposited at bath temperature of: (a) T_b = 70°C, (b) T_b = 75°C, (c) T_b = 80°C, (d) T_b = 85°C and (e) T_b = 90°C, with deposition time of 60 min.

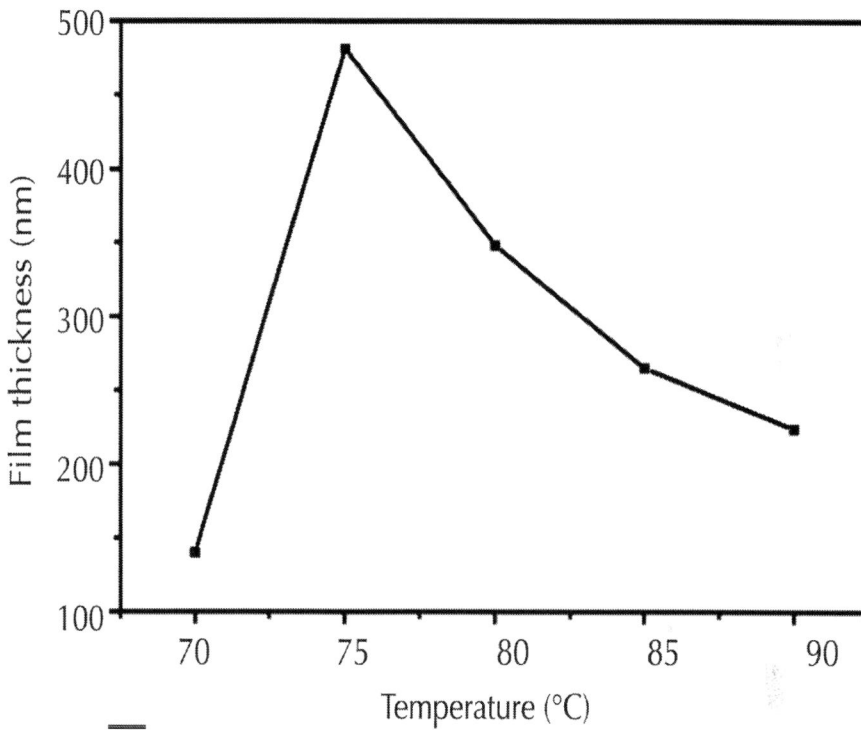

Figure 2: Variation of CdS film thickness with bath temperature at t_d = 60 min.

Figure 3 shows the X-Ray diffraction diagrams of the CdS thin films prepared with different deposition times at bath temperature of 75°C. As it can be seen the intensity of the (002) plane increases with increasing deposition time (as above 60 min) and then decreases when deposition time increases. This indicates that 60 min is the optimum deposition time for which the crystallinity of the films reaches the maximum. Figure 4 depicts the variation of the films thickness in function of deposition time. Two different regions can be observed, a first region, where the film thickness varies approximately linearly with deposition time up to 60 min, indicating a constant growth, and a second region, where a film growth reaches saturation and decreases when the deposition time is prolonged. The same behaviour is also observed for other chalcogenide materials such as ZnS [13] and MnS [19]. Lokhande et al. [16] have explained this phenomenon by considering two competing processes taking place in the deposition bath: one process includes the heterogeneous and homogenous precipitation of CdS, which leads to the film growth; the other one involves the dissolution of the pre-formed CdS film, which result in the decrease of film thickness. In the initial time of deposition, the source materials are sufficient; the process of heterogeneous and homogenous precipitation play a more important role than the dissolution process, leading to the increase in film thickness. When the deposition time is prolonged (60 min in our case), the source materials become less. Therefore, the dissolution process predominates over the heterogeneous and homogeneous precipitation, resulting in the decrease of film thickness.

Table 1: Average crystallite size and thickness of the CdS thin films prepared at different bath temperatures

Bath temperature (SC)	Average crystallite size (nm)	Film thickness (nm)
70	11.80	140
75	20.87	481
80	19.11	348.2
85	15.11	265

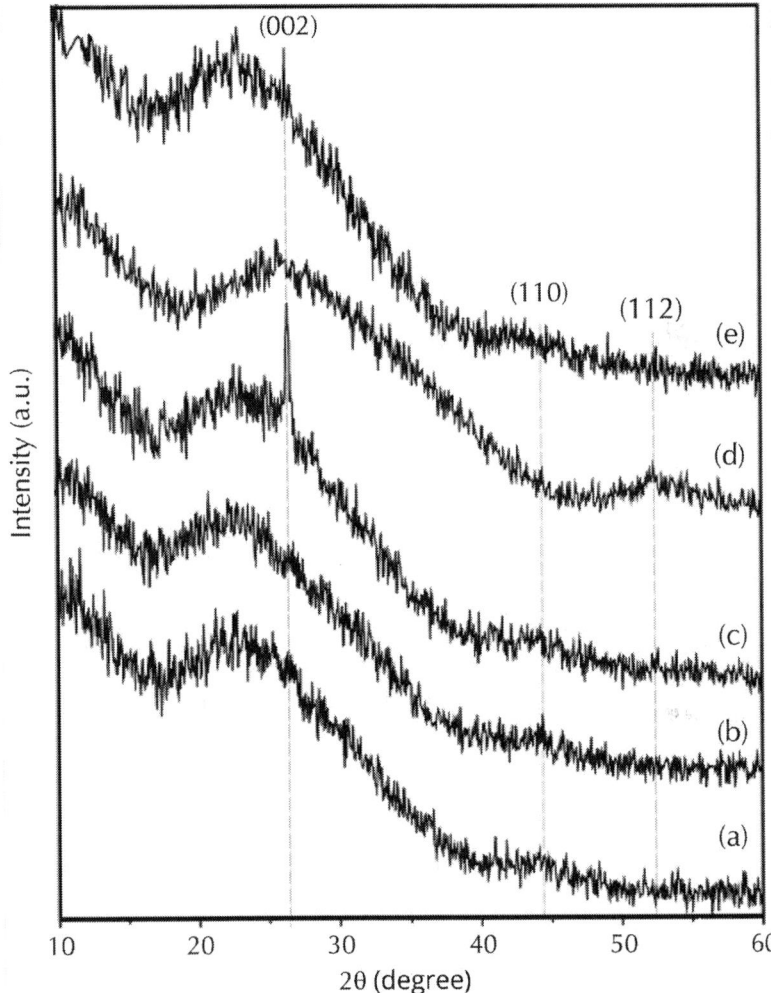

Figure 3: X-ray diffraction patterns of CdS thin films prepared with different deposition times: (a) t_d = 15 min, (b) t_d = 30 min, (c) t_d = 60 min, (d) t_d = 90 min and (e) t_d = 120 min. T_b = 75°C.

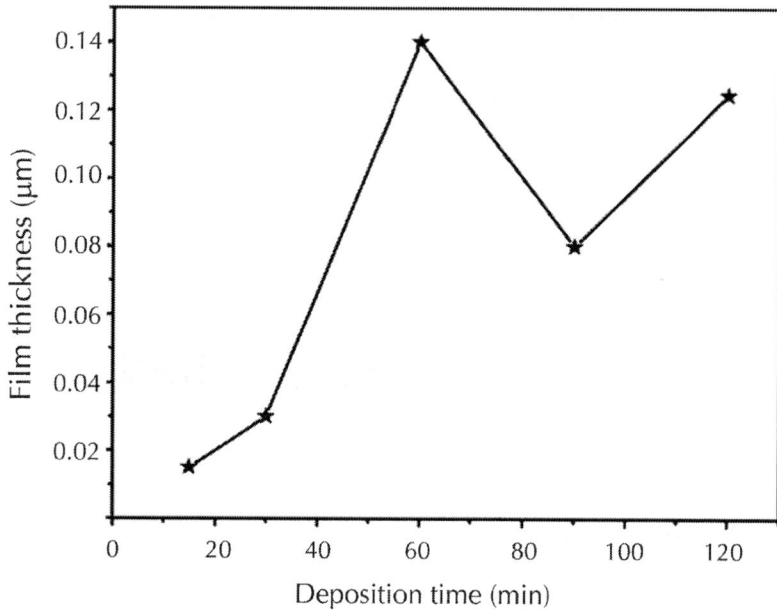

Figure 4: Variation of CdS film thickness with deposition time at T_b = 75˚C.

Figure 5 shows the X-Ray diffraction diagrams of the CdS thin films prepared with different [S]/ [Cd] ratios. The intensity of the (002) plane is found to be increased when increasing [S]/ [Cd] ratio up to 2.5, which decreased afterwards. The decrease of the crystallinity for the films deposited with [S]/ [Cd] ratio greater than 2.5 is attributed to the decrease in films thickness. Figure 6presents the variation of the films thickness with the [S]/ [Cd] ratio in the solution. As it can be seen, the thickness reaches the maximum when the films are deposited with [S]/[Cd] ratio equals to 2.5, this is agree with the R. Mendoza-Pérez et al. work [20]. The decrease of the films thickness for films prepared with [S]/[Cd] ratio greater than 2.5 is probably due to the dissolution process of the preformed CdS films and the desorption phenomenon, which are predominates over the heterogeneous and homogeneous process.

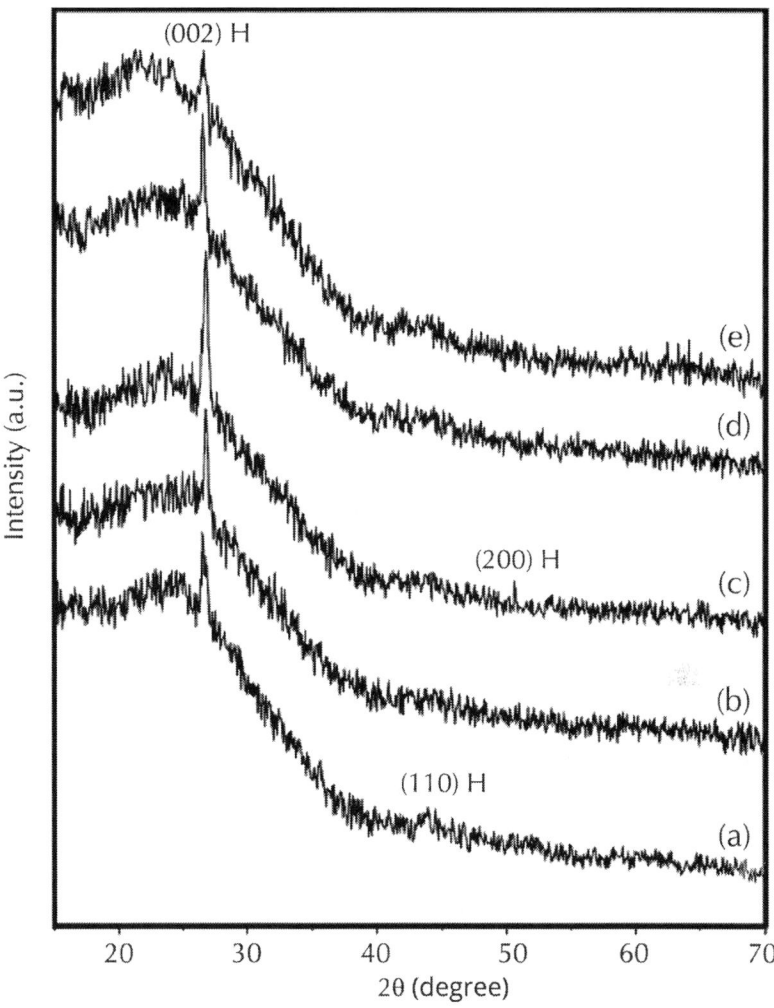

Figure 5: X-ray diffraction patterns of CdS thin films elaborated with different [S]/[Cd] ratios: (a) [S]/[Cd] = 1, (b) [S]/[Cd] = 2, (c) [S]/[Cd] = 2.5, (d) [S]/[Cd] = 3 and (e) [S]/[Cd] = 5.

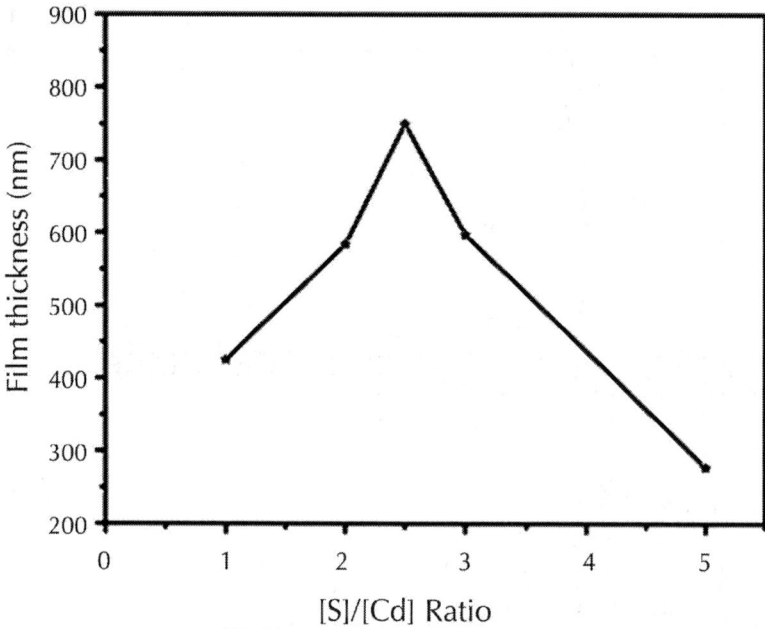

Figure 6: Variation of CdS film thickness with [S]/[Cd] ratio. $t_d = 60$ min, $T_b = 75°C$.

Surface Morphology

The influence of the temperature on the morphology of the CdS films is shown in Figure 7. It is clearly seen that the average crystallite size decreases with increasing temperature. For T = 90°C, no grains are observed indicating an amorphous phase for films prepared at this temperature. These results are in agreement with those obtained by X-ray diffraction (Table 1).

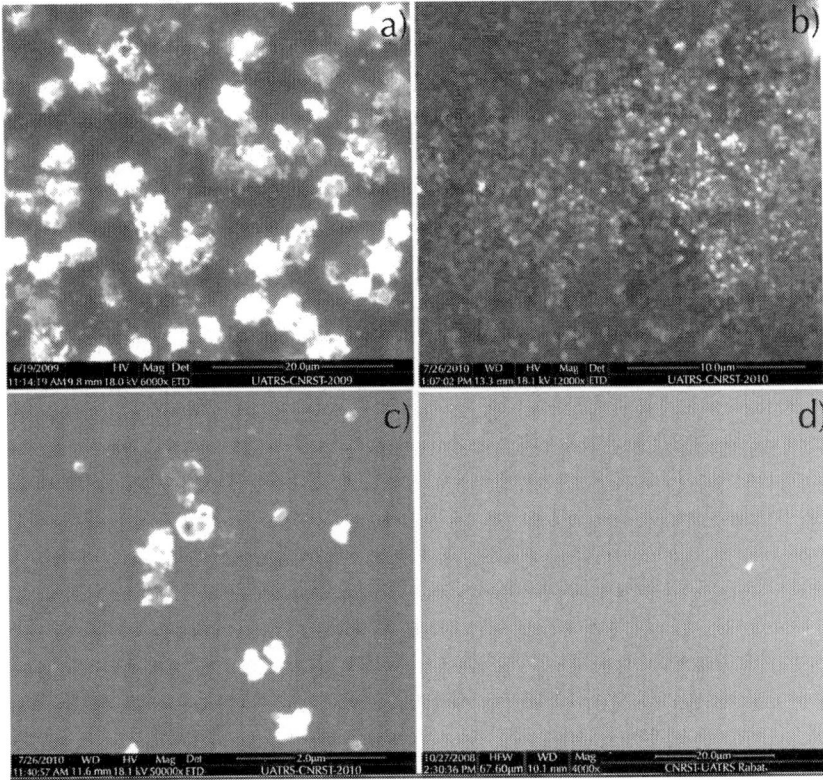

Figure 7: SEM micrographs of CdS thin films deposited at bath temperature of: (a) T_b = 75°C, (b) T_b = 80°C, (c) T_b = 85°C and (d) T_b = 90°C.

The SEM micrographs of the surface morphology of the CdS films prepared with different [S]/[Cd] ratios in the solution (1, 2, 2.5, 3 and 5) are shown in Figure 8. These micrographs show that the obtained films have good adherence on the substrates without pinholes or cracks. Moreover, The films are covered by spherical grains, whose size decreases and their density increases noticeably when [S]/[Cd] ratio increases. When [S]/[Cd] ratio is 1 (Figure 8(a)), the small particles accumulate continuously and cover the entire surface of the substrate leading to an homogeneous layer. This indicates that the mechanism film formation is due to ion-by-ion deposition (heterogeneous mechanism). When [S]/[Cd] ratio increase up to 1, a small particles are grouped to form larger clusters discreetly distributed in the films as it is shown clearly in Figures 8(b)-(e). This indicates that the mechanism film formation is due to cluster-by-cluster deposition (homogeneous mechanism). The average crystallite size of these films varies between 500 and 1000 nm while that estimated by Scherrer's equation is 21 nm. This value is much less than that from SEM. This is can be explained by agglomeration of small particles of the CdS to form large clusters. To have more details on the surface morphology of the CdS films, the AFM analysis is used.

Figure 9 presents 3D AFM images obtained by scanning an area of 2 µm × 2 µm (Figure 9(a)), 20 µm × 20 µm (Figure 9(b)) and 20 µm × 20 µm (Figure 9(c)) of the surface of the CdS films deposited with ratio [S]/[Cd] = 2.5 at 75°C and deposition time of 30 min, 75°C and deposition time of 60 min and 85°C and deposition time of 60 min, respectively. They show that the surfaces are composed of small grains with an approximate size of 20 to 30 nm and grains grouped together to form large clusters like a cauliflower with a mean size of 200 to 500 nm, confirming the results obtained by X-ray diffraction analysis and SEM, respectively. Moreover, the film deposited with a short deposition time (30 min) (Figure 9(a)) shows a strong surface roughness (50 to 60 nm) compared to that of film (30 to 40 nm) prepared with a long deposition time (60 min) (Figure 9(b)), suggesting that, at the beginning, the mechanism CdS films formation is probably due to clusters-by-clusters deposition (homogeneous mechanism). On the other hand, when the bath temperature is reached to 85°C, none a significant improvement in the surface roughness is noticed.

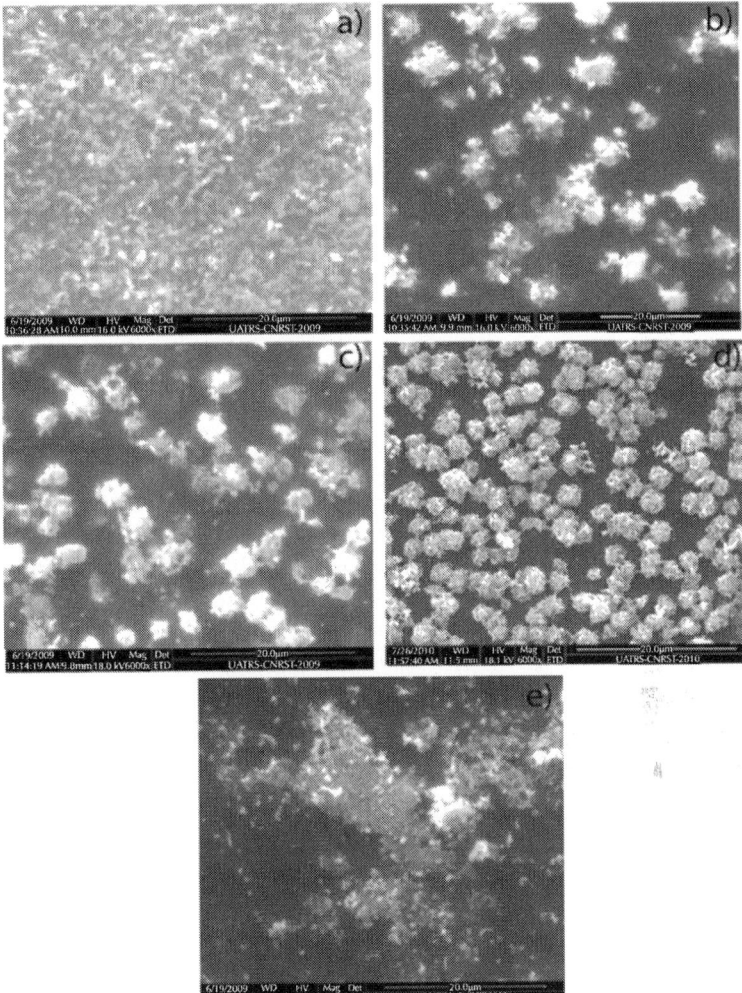

Figure 8: SEM micrographs of CdS thin films deposited with different [S]/[Cd] ratios: (a) [S]/[Cd] = 1, (b) [S]/[Cd] = 2, (c) [S]/[Cd] = 2.5, (d) [S]/[Cd] = 3 and (e) [S]/[Cd] = 5.

Composition Analysis

Figure 10 presents the EDAX spectrum of the CdS thin films prepared under optimum growth conditions. It shows peaks of Cd, S and some impurities like, Si, Ca, Na and O which are originated probably from the glass substrates and deionised water, respectively. The atomic concentrations from EDAX analysis and the calculated atomic ratio are presented in Table 2. As it can be seen, the [S]/[Cd] atomic ratio is 0.94 suggesting the presence of sulphur vacancies (excess of cadmium) in the deposited films, which act as donors, leading to n-type conductivity.

Optical Properties

The optical properties such as transmittance, absorption coefficient and band gap energy of CdS thin films are determined from the variation of the optical transmission with wavelength ( ) in the range of 300 to 1500 nm (Table 1). Figure 11 shows plots of optical transmission of the CdS films deposited at different bath temperatures. The films produced at optimum conditions (T_b = 75°C) are found relatively to be highly transparent of about 80% in the visible and near infra-red regions. This is can be explained by a less light scattering of these films due to their smoothest surfaces. The similar phenomenon has been reported in the literature [21, 22].

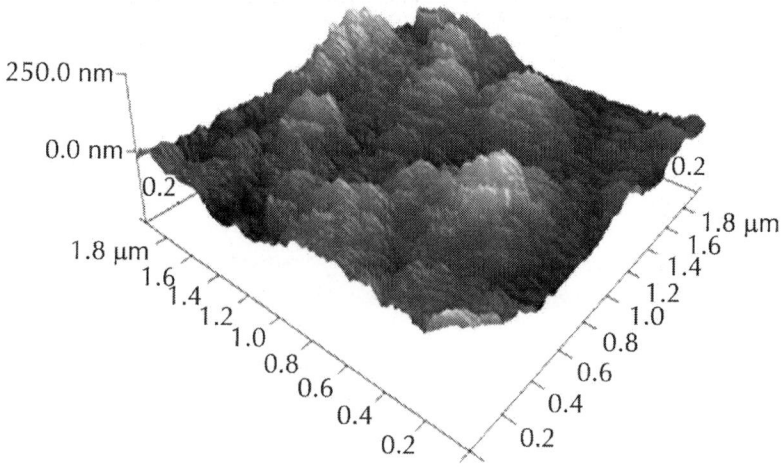

(a)

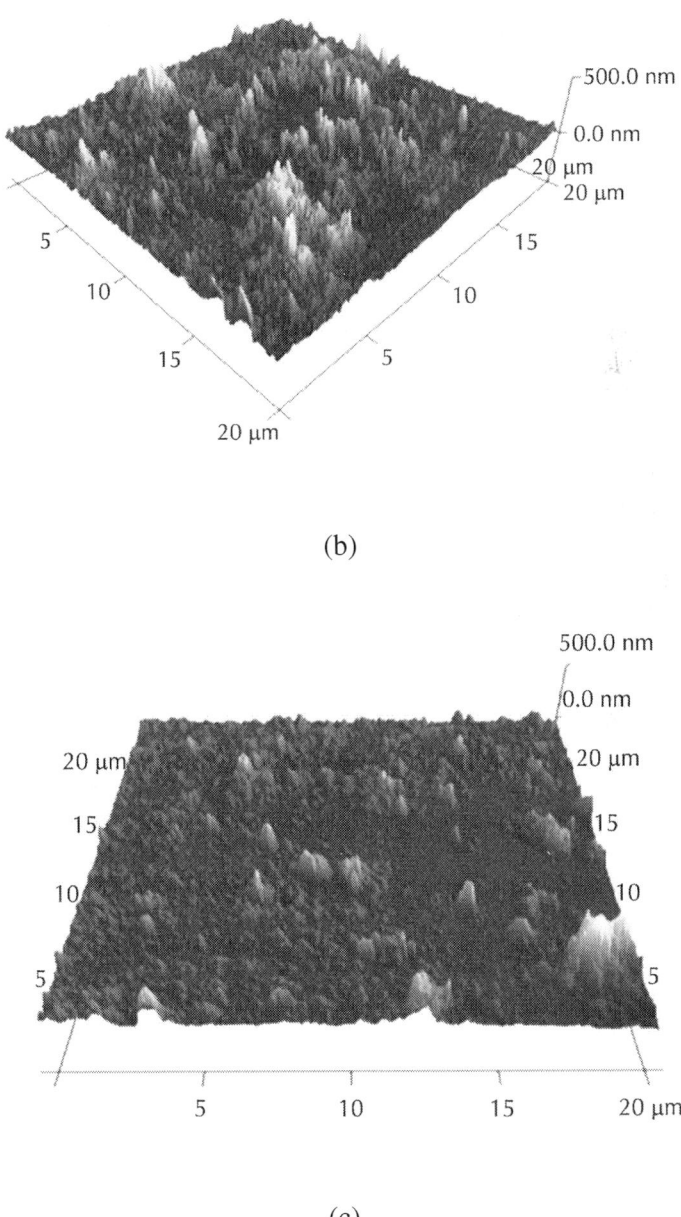

(b)

(c)

Figure 9: 3D AFM micrographs of CdS thin films deposited with [S]/[Cd = 2.5 at: (a) T_b = 75°C and t_d 30 min, (b) T_b = 75°C and t_d = 60 min, (c) T_b = 85°C and t_d = 60 min.

By using the Tauc relationship which is given by the formula [23]:

$$(\alpha h v) = A(h v - Eg)^{n}$$

(7)

In witch, h is the photon energy, Eg is the optical band gap of the semiconductor, A is a constant and n = 1/2 for direct band gap semiconductor such as CdS, the optical band gap value of the CdS thin films is estimated by extrapolation of the straight line of the plot of (h)^2 versus photon energy as it is shown in Figure 12. We found that the value of Eg for the films prepared at different temperatures varies from 2.40 to 2.46 eV, which is in agreement with the value reported by other authors [21,24].

Table 2: EDAX analysis of the CdS thin films deposited under optimum conditions: T_b = 75°C, t_d = 60 min and [S]/ [Cd] = 2.5

Element	Cd	S	O	[S]/[Cd]
Percentage in at %	7.6	7.2	37.7	0.94

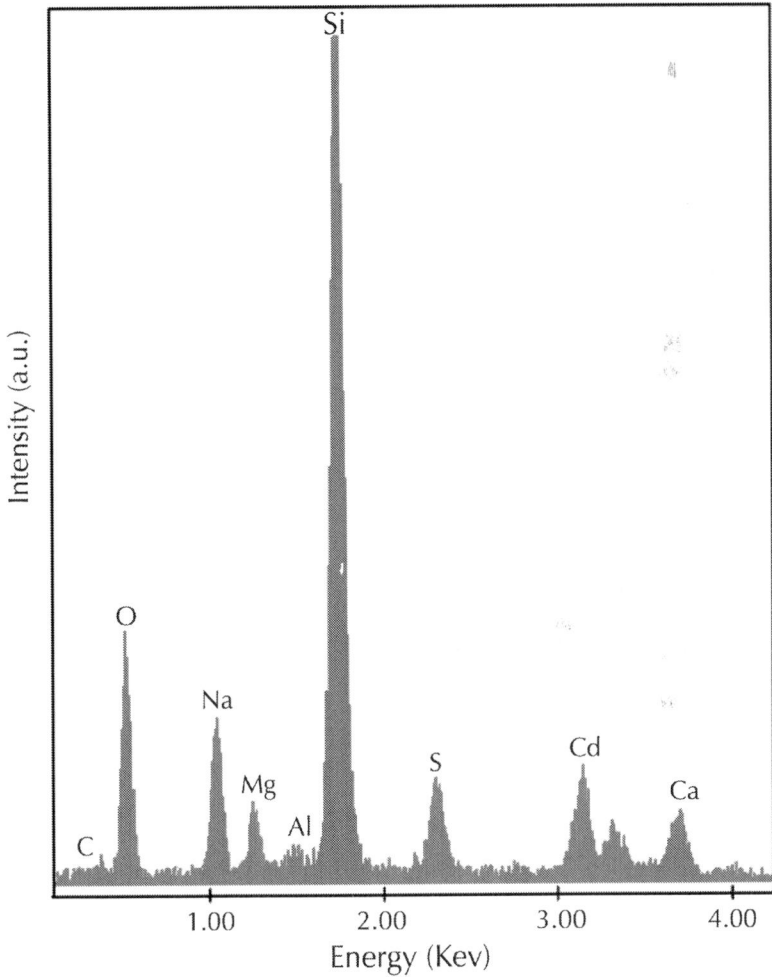

Figure 10: EDAX spectrum of CdS thin film deposited under optimum conditions: $T_b = 75°C$, $t_d = 60$ min and $[S]/[Cd] = 2.5$.

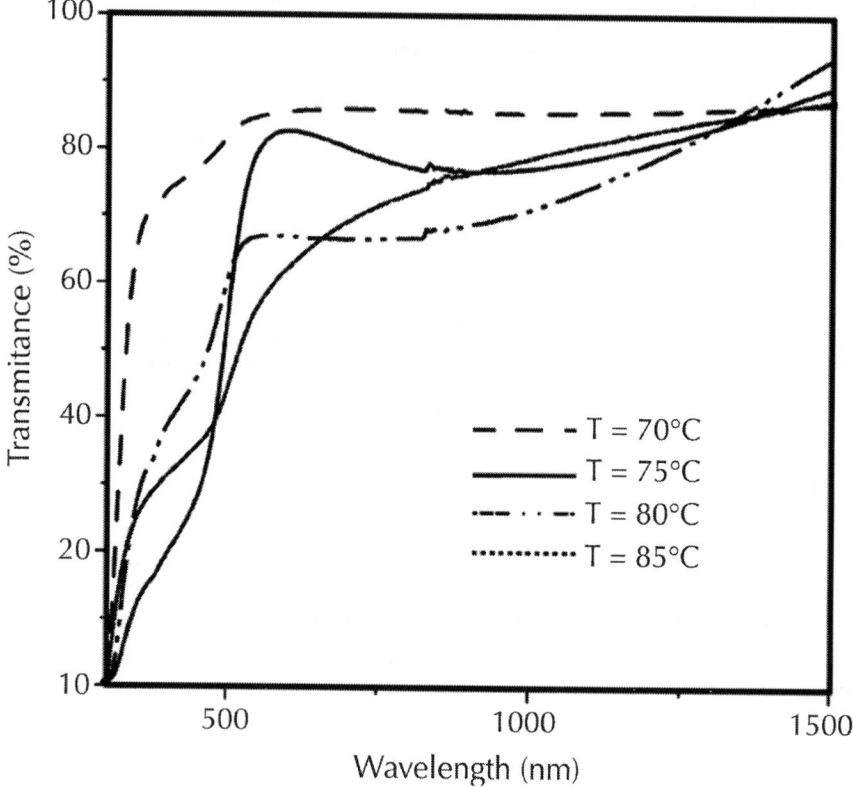

Figure 11: transmission spectra of CdS thin films prepared at different bath temperatures.

The effect of [S]/[Cd] ratio on the optical properties of CdS films deposited at T = 75°C and t_d = 60 min is studied. Figure 13 shows the optical transmittance spectra of the CdS films elaborated with different [S]/[Cd] ratios. As it can be seen, the films elaborated with [S]/[Cd] ratio equals 2.5 exhibit a high transmission of about 80% in the visible and near infra-red regions. This can be explained relatively by a good cristallinity and stiochiometry of these films as it is found by X-ray diffraction and EDAX analysis. The optical band gap of these films is also calculated by extrapolation of the straight line of the plot of (h)^2 versus photon energy as it is shown in Figure 14. The values of Eg, average transmittance (%T) and average crystallite size (D_{hkl}), are compiled in Table 3. It can be observed that the CdS films prepared with [S]/[Cd] = 2.5 have high average transmission of about 80% in the visible and infrared regions, large band gap of 2.44 eV, which is close to the Eg value reported for single crystalline CdS (2.43 eV) and relatively large grain size.

CONCLUSIONS

CdS thin films were deposited on glass substrates by the CBD method using a solution of cadmium chloride and thiourea as sources of cadmium and sulphur ions, respectively. The optimal deposition parameters were found to be T_b = 75˚C, t_d = 60 min and [S]/[Cd] = 2.5. The films prepared under these optimal conditions were relatively well crystallized and had hexagonal structure with a preferential orientation along the (002) direction. These films showed also large final thickness and their surface morphologies were composed of small grains with an approximate size of 15 to 25 nm and grains grouped together to form large clusters like a cauliflower. The composition study showed that these films were nonstoichiometric with a slight sulphur deficiency leading to n type conductivity. These films exhibited also a good transmittance of about 80% in visible and near infra-red regions of the electromagnetic spectrum so it is possible to us them as a window layer in high efficiency thin film solar cells based on CdTe and $Cu(In,Ga)Se_2(CIGS)$.

ACKNOWLEDGEMENTS

The authors are gratefully to "Platform for Imaging and Measurement in Microscopy of University of Brest, France" for assistance in AFM measurements.

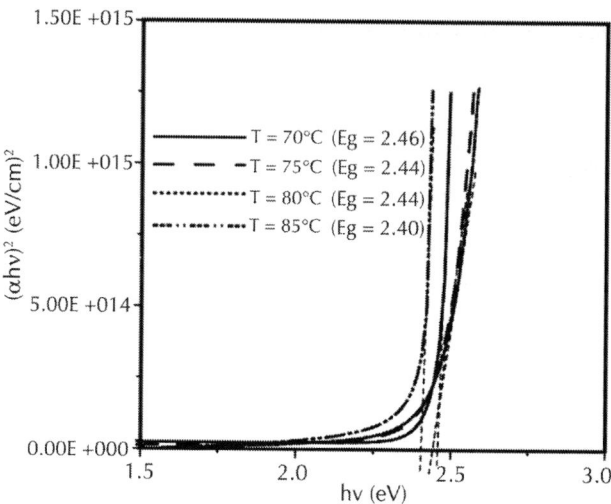

Figure 12: Variation of $(\alpha h\nu)^2$ with photon energy (hν) for CdS thin films elaborated at different bath temperatures.

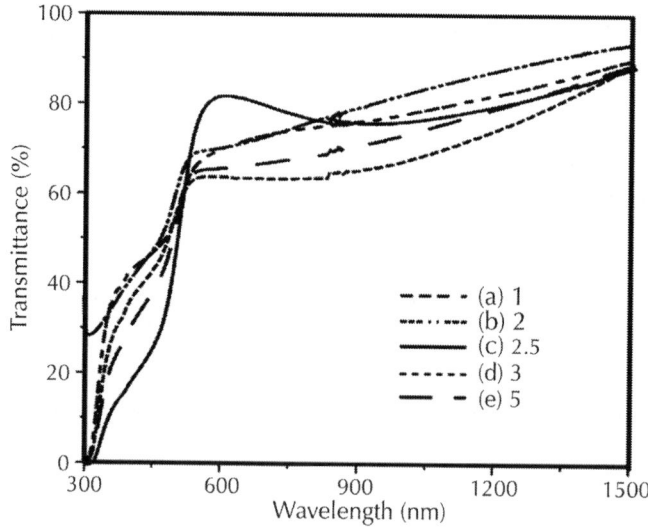

Figure 13: Optical transmission spectra of CdS thin films prepared with different [S]/[Cd] ratios.

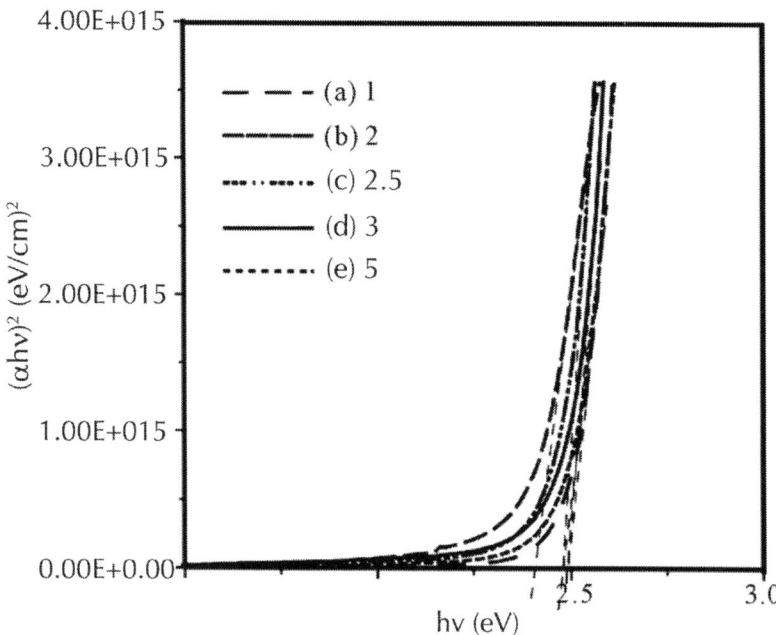

Figure 14: Variation of $(\alpha h\nu)^2$ with photon energy (hv) for CdS thin films elaborated with different [S]/[Cd] ratios.

Table 3: Average crystallite size (D_{hkl}), band gap (Eg) and average transmittance values (%T) of the CdS thin films with varying ratios of [S]/[Cd]

[S]/[Cd]	Eg (eV)	Average transmittance (%)	Average crystallite size (nm)
1	2.37	69	11.84
2	2.39	71	16.51
2.5	2.44	79	20.87
3	2.42	75	18.34
5	2.36	66	9.90

REFERENCES

1. M. Contreras, M. Romero, B. To, F. Hasoon, R. Noufi, S. Ward and K. Ramanathan, "Optimization of CBD CdS Process in High-Efficiency Cu(In,Ga)Se2-Based Solar Cells," Thin Solid Films, Vol. 403-404, No. 579, 2002, pp. 204-211.

2. A. Davis, K. Vaccaro, H. Dauplaise, W. Waters and J. Lorenzo, "Optimization of Chemical Bath-Deposited Cadmium Sulfide on InP Using a Novel Sulfur Pretreatment," Journal of The Electrochemical Society, Vol. 146, No. 3, 1999, pp. 1046-1053.

3. O. Vigil-Galan, J. Ximello-Quiebras, J. Aguilar-Hernandez, G. Contreras-Puente, A. Cruz-Orea, J. Mendoza-Alvarez, J. Cardona-Bedoya, C. Ruiz and V. Bermudez, "Passivation Properties of Cds Thin Films Grown by Chemical Bath Deposition on Gasb: The Influence of the S/Cd Ratio in the Solution and of the Cds Layer Thickness on the Surface Recombination Velocity," Semiconductor Science and Technology, Vol. 21, 2006, pp. 76.

4. M. Ilieva, D. Dimova-Malinovska, B. Ranguelov and I. Markov, "High Temperature Electrodeposition of Cds Thin Films on Conductive Glass Substrates," Journal of Physics: Condensed Matter, Vol. 11, No. 49, 1999, pp. 10025-10031.

5. A. Aschour, "Physical Properties of Spray Pyrolysed CdS Thin Films," Turkish Journal of Physics, Vol. 17, No. 8, 2003, pp. 551-558.

6. B. Pradhan, A. K. Sharma and A. K. Ray, "Conduction Studies on Chemical Bath Deposition Nanocrystalline Cds Thin Films," Journal of Crystal Growth, Vol. 304, No. 2, 2007, pp. 388-392.

7. P. Boieriu, R. Sporken, Y. Xin, N. Browning and S. Sivananthan, "Wurtzite CdS on CdTe Grown by Molecular Beam Epitaxy," Journal of Electronic Materials, Vol. 29, No. 6, 2000, pp. 718-722.

8. H. Uda, H. Yonezawa, Y. Ohtsubo, M. Kosaka and H. Sonomura, "Thin CdS Films Prepared by Metalorganic Chemical Vapor Deposition," Solar Energy Materials and Solar Cells, Vol. 75, 2003, pp. 219-226.

9. M. Sasagawa and Y. Nosaka, "The Effect of Chelating Reagents on the Layer-by-Layer Formation of Cds Films in the Electroless and

Electrochemical Deposition Processes," Electrochimica Acta, Vol. 48, No.5, 2003, pp. 483-488.

10. R. W. Birkmire, B. E. McCandless and S. S. Hegedus, "Optimization of Vapor Post-Deposition Processing for Evaporated CdS/CdTe Solar Cells," Solar Energy, Vol. 12, 1992, pp. 37-45.

11. G. Hodes, "Chemical Solution Deposition of Semiconductors Films," Marcel Dekker, Inc., Basel, New York, 2003.

12. ASTM DATA (6-0314).

13. R. Zhai, S. Wang, H. Xu, H. Wang and H. Yan, "Rapid Formation of CdS, ZnS Thin Films by MicrowaveAssisted Chemical Bath Deposition," Matarials Letters, Vol. 59, No. 12, 2005, pp. 1497-1501.

14. O. Oladeji, L. Chow, J. R. Liu, W. K. Chu, A. N. P. Bustamante, C. Fredricksen and A. F. Schulte, "Comparative Study of Cds Thin Films Deposited by Single, Continuous, and Multiple Dip Chemical Processes," Thin Solid Films, Vol. 359, No. 2, 2000, pp. 154-159.

15. M. Ichimura, F. Goto and E. Arai, "Structural and Optical Characterization of Cds Films Grown by Photochemical Deposition," Journal of Applied Physics, Vol. 85, No. 10, 1999, pp. 7411-7417.

16. C. D. Lokhande, A. Ennaoui, P. S. Patil, M. Giersig, M. Muller, K. Diesner and H. Tribursch, "Process and Characterisation of Chemical Bath Deposited Manganese Sulphide (MnS) Thin Films," Thin Solid Films, Vol. 330, No. 2, 1998, pp. 70-75.

17. S. Prabahar and M. Dhanam, "CdS Thin Films from Two Different Chemical Baths-Structural and Optical Analysis," Journal of Crystal Growth, Vol. 285, No. 1-2, 2005, pp. 41-48.

18. L. D. Kadam and P. S. Patil, "Thickness-Dependent Properties of Sprayed Cobalt Oxide Thin Films," Materials Chemistry and Physics, Vol. 68, No. 1-3, 2001, pp. 225-232.

19. D. Fan, H. Wang, Y. Zhang, J. Cheng. B. Wong. H. Yan, "Preparation of Crystalline MnS Thin Films by Chemical Bath Deposition," Materials Chemistry and Physics, Vol. 80, 2003, pp. 44-47.

20. R. Mendoza-Pérez, G. Santana-Rodriguez, J. Sastre-Hernandez, A. Morales-Acevedo, A. Arias-Carbajal, O. VigilGalan, J. C. Alonso and G. Contreras-Puente, "Effects of Thiourea Concentration on

CdS Thin Films Grown by Chemical Bath Deposition for CdTe Solar Cells," Thin Solid Films, Vol. 480-481, 2005, pp. 173- 176.

21.   F. Y. Liu, Y. Q. Lai, J. Liu, B. Wang, S. S. Kuang, Z. A. Zhang, J. Li and Y. X. Liu, "Characterization of Chemical Bath Deposited CdS Thin Films at Different Deposition Temperature," Journal of Alloys and Compounds, Vol. 493, No. 1-2, 2010, pp. 305-308.

22.   K. S. Ramaiah, R. D. Pilkington, A. E. Hill, R. D. Tomlinson and A. K. Bhatnagar, "Structural and Optical Investigations on Cds Thin Films Grown by Chemical Bath Technique," Materials Chemistry and Physics, Vol. 68, No. 1-3, 2001, pp. 22-30.

23.   J. Tauc, "Amorphous and Liquid Semiconductors," Plenum Press, New York, 1974, pp. 159-220.

24.   L. Wenyi, C. Xun, C. Qiulong and Z. Zhibin, "Influence of Growth Process on the Structural, Optical and Electrical Properties of CBD-Cds Films," Materials Letters, Vol. 59, No. 1, 2005, pp. 1-5.

Use of Poly-Lactic Acid (PLA) to Enhance Properties of Paper Based on Recycled Pulp

Klaus Doelle, Anh T. Le, Thomas E. Amidon, and Biljana M. Bujanovic

Department of Paper and Bioprocess Engineering, State University of New York College of Environmental Science and Forestry, Syracuse, USA

ABSTRACT

Nowadays, recycled paper is broadly used due to environmental reasons. Furthermore, the addition of starch as a dry strength additive improves the properties of recycled paper. Poly-Lactic Acid (PLA), a product from bio-refinery process, has recently been shown to act as a promising strength additive that could be used in combination with starch to further improve the strength of paper. In this study, the use of PLA of three molecular weights (MW) in combination with four different starches was investigated. Three recycled pulps from different origins, with the kappa number of 27.9 to 66 were used. Paper handsheets were made, and selected paper properties were tested. The results indicate that handsheets properties were influenced by the MW of PLAs, the type of starch used, and the lignin content of the pulp. The paper handsheets made from lignin-rich pulp (pulp A, kappa number 66), combined with 0.1% medium MW PLA (PLA_1) and 0.9% cationic starch containing 0.43% N gave the highest improvement for tensile strength, wet tensile strength, air and water resistance. This result verifies that a higher kappa number pulp has better attraction to the hydrophobic PLA. Moreover, the higher charge cationic starch led to higher tensile strength due to the increase of affinity to the anionic fiber surface. Interestingly, results show that amphoteric starch is a promising substitute for high cationic charge starch when combined with the medium MW PLA to improve tensile strength of paper. This study demonstrated that a starch-PLA blend represents a promising approach in improving properties of recycled paper.

INTRODUCTION

Currently, papermakers are focusing on how to improve strength for paper. This can be measured by tensile strength, bursting strength, or internal bonding strength. There are various ways to increase the strength of paper such as using refining, wet pressure, or additives. However, use of mechanical actions like refining could bring about the reduction in opacity, brightness, dimensional stability or porosity [1] . Therefore, use of additives is very common and has been employed in most paper mills. Starch is used most frequently today as a dry strength additive in the paper industry. Starches can significantly improve the mechanical properties of paper such as tensile strength. Also, starches are used as retention aid, surface sizing agent, coating binder, and adhesive in corrugated board and other converting operations [2] .

Beside the excellent contributions to the strength and other properties of paper, the use of starches can lead to several difficulties for papermaking process. One of the current major problems when using starches is the loss of starch to the white water system. Starches are water soluble, hydrophilic polymers [3] . When the amount of starch added to the system is high, the retention of starch on the paper web becomes lower and then starch would be lost to the white water system due to its hydrophilic properties. The loss of starches results in increasing production costs, and in the long term of operation, it can cause microbiological problems that negatively affect the runnability of machines as well as the quality of paper. Moreover, because starch is mainly considered as a dry wet strength agent, it is not known to be a contributor to the wet strength, air and water resistance of paper. Therefore, it is necessary to study new additives to use together with starch to further increase the strength of paper, reduce the consumption of starch, reduce the microbiological problems, and improve wet strength, air and water resistance of paper.

In an attempt to fulfil these needs, poly-lactic acid (PLA) had been introduced to the wet end system. PLA has been found to be a potential alternative for starch, and it is also a biodegradable product. Furthermore, PLA can be produced from natural resources like corn, but also woody biomass. Recently PLA has been shown in the laboratory study to have the function of a strength additive [4] [5] . The laboratory result was reported that the blend of PLA with cationic starch as a wet end additive improved the tensile strength of paper. Regarding to Gong, several virgin pulps had been used including Norway spruce thermo-mechanical pulp (TMP), hardwood unbleached kraft pulp, and softwood bleached kraft pulp. Two cationic starches that have different nitrogen contents, and PLA with molecule weight of 20,000 - 30,000 were employed.

In this research, three types of recycled pulp were used instead of the virgin pulps. This was done because recycled paper has been used broadly currently due to environmental impact. Furthermore, it has been shown that some paper properties from recycled pulp could be comparable to that of virgin pulp. For example, after several times of recycling, mechanical pulp still gives better density, better burst strength and better tensile strength compared to the never-dried mechanical pulp [6] . However, for chemical pulp, important paper properties like burst and tensile strength decrease as the number of recycles increase. The reason for the decrease in strength is due to the loss of bonding potential between fibers after the fibers were dried in the paper drying process, commonly known as "hornification" phenomenon. One of the methods to recover the strength of paper is to use the additives such as modified starch.

In this present paper, we report the properties of paper handsheets made from the three recycled pulps, combined with three types of PLA and four different starches.

EXPERIMENTAL

Materials

Recycled pulp (Pulp A) was donated by a paper company in Central New York. Kappa number of this pulp was 66. The initial Canadian Standard Freeness (CSF) was 606 ml. The pulp was then refined to CSF of 305 ml which is close to CSF of 300 ml. Regarding TAPPI (Technical Association of Pulp and Paper Industry) T 200, CSF of 300 ml is recommended because many pulps show a maximum strength at this freeness. Recycled pulp (Pulp B) was supplied by a paper company from Vietnam. Kappa number was found at 27.9 and the initial CSF was found at 419 ml. The pulp was refined to a CSF of 286 ml. Recycled pulp (Pulp C) was the same as Pulp A but it was refined to a different CSF, 298 ml. Linerboard samples were obtained from a paper mill in Central New York. The linerboard was produced from 100% recycled fibers, and was treated with the internal sizing agent alkyl ketene dimer (AKD).

Cationic starch 1 (CAT.N1) is a commercial product and was supplied by a company in the United States. The nitrogen content of this starch is 0.28%. Cationic starch 2 (CAT.N2) with 0.43% nitrogen content, was supplied by a different chemical company. Amphoteric starch (AS), a commercial product, was provided by the same company as for the CAT.N1. The nitrogen and phosphate content were 0.28% and minimum 1%, respectively. Anionic starch (ANS), a commercial product, was obtained from the same company as the CAT.N1.

There are three different types of Poly-lactic acid (PLA) employed. They were all purchased from Polysciences, Inc in the United States. Poly-lactic acid 1 (PLA_1) is poly (dl-lactic acid) with a molecular weight MW ~20,000 - 30,000. Poly-lactic acid 2 (PLA_2) is poly (l-lactic acid) with a molecular weight MW ~1600 - 2400. Poly-lactic acid 3 (PLA_3) is poly (l-lactic acid) with a molecular weight MW ~140,000 - 160,000.

Methods

TAPPI Methods

Kappa number of the recycled pulp was measured in accordance with TAPPI T 236 om-06, "Kappa number of pulp". The CSF was measured by T 227 om-09 "Freeness of pulp (Canadian standard method)". Paper handsheets were made followed T 205 sp-06, "Forming handsheets for physical tests of pulp". Paper specimens were prepared using the T220 sp-10, "Physical testing of pulp handsheets". Grammage and Thickness of handsheets were determined by the T220 sp-10. The tensile strength was performed following T 494 om-06, "Tensile properties of paper and paperboard (using constant rate of elongation apparatus)". The wet tensile strength was measured on the same machine used for dry tensile strength, and the method used was T 456 om-10, "Tensile breaking strength of water-saturated paper and paperboard ("Wet tensile strength"). The Gurley porosity was performed by following the T 460 om-11, "Air resistance of paper (Gurley method)". The tear strength was done by following the T 414 om-12, "Internal tearing resistance of paper (Elmendorf-type method)". The Cobb values were obtained by the T441 om-09, "Water absorptiveness of sized (non-bibulous) paper, paperboard, and corrugated fiberboard (Cobb test)". Water drop was measured by T 835 om-08, "Water absorption of corrugating medium: water drop absorption test".

PLA Treatment

Because PLA is not soluble in water, it was dissolved in organic solvents. Furthermore, because different molecular weights of PLA were used, two different solvents were used. PLA_1 was dissolved in acetone easily under normal stirring condition and a short period of time. PLA_1 can be dissolved completely in a minimal amount of acetone. In this study, 0.02 g of PLA_1 was completely dissolved in 20 ml acetone. PLA_2 was dissolved in acetone but under long and high stirring condition. PLA_3 was dissolved in p-dioxane under long and high stirring condition.

Starch Preparation

Starch in the form of powder was mixed with distilled water and then cooked at the consistency 0.1% at temperature 95°C - 98°C for 30 minutes under constant stirring. During the cooking, a flask containing the starch solution was covered by a small aluminium tray to prevent the evaporation of water.

Mixing of Starch-PLA

After PLA is dissolved in acetone, it was mixed with the cooked starch and they together formed a clear and stable solution. Then the starch-PLA mixture was added to the pulp slurry and agitated for about 5 minutes to get an even mixture of pulp, starch and PLA.

RESULTS AND DISCUSSION

Selected properties of paper handsheets made from the three different pulps, from the four different starches and from the three different PLAs are presented as following discussions. Table 1is introduced to show the repeatability of different properties. This table will be used for comparison purpose in the following discussions.

Effect of CAT.N1 on Paper Properties of Pulps A and B

Effect on Strength Properties

Figure 1 shows the effect of this starch on the strength properties of paper made from the two different pulps, Pulp A and Pulp B.

For Tensile Index (TI), Figure 1 shows that for blank handsheet, the TI of paper made from Pulp A (33.10 Nm/g) is greater than that of paper made from Pulp B (27.40 Nm/g). When 1% CAT.N1 is used, TI of paper made from Pulp A increases rapidly to 45.60 Nm/g (~37.8%), while the TI of paper made from Pulp B increases to 32.70 Nm/g (19.3%). When 0.1% PLA_1 is introduced, TI of paper made from

Pulp A decreases to 42.50 Nm/g while TI of paper made from Pulp B slightly increases to 33.70 Nm/g. With the introduction of PLA_2, TI of both paper made from Pulp A and Pulp B become lower to 38.40 and 32.30, respectively. Therefore, the results propose that CAT.N1 has better effect on TI of Pulp A at 1%. And CAT.N1 has better effect on TI of Pulp B at 0.9% combining with 0.1% PLA_1.

Table 1: Repeatability of test results

Tests	Repeatability	Recommended by TAPPI
Tensile Strength (TS)	5%	T494
Wet Tensile Strength (WTS)	4%	T456
Tear	4%	T414
Gurley Porosity	8%	T460
Water Drop	13%	T835
Cobb	2.5%	T441

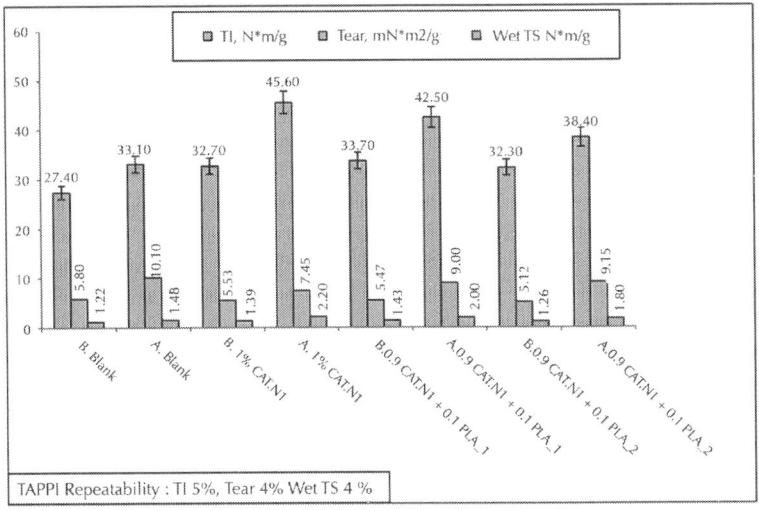

Figure 1: Effects of CAT.N1 on strength properties of Pulps A and B.

Similar to TI, with the use of 1% CAT.N1 alone paper made from pulp A shows higher wet TS than paper from Pulp B (2.20 vs. 1.39 Nm/g). However, when combining with 0.1% PLA, wet TS of paper from Pulp B slightly increases from 1.39 to 1.43 Nm/g while it decreases from 2.20 to 2.00 Nm/g for paper made from Pulp A.

Tear values decrease for both Pulp A and Pulp B compared to the blank experiment in all cases. Interestingly, for Pulp A, the use of PLA_1 and PLA_2 with the CAT.N1 leads to the increase of the tear index compared to the use of 1% CAT.N1 only. Particularly, tear index of Pulp A + 0.9 CAT.N1 + 0.1 PLA_1 and Pulp A + 0.9 CAT.N1 + 0.1 PLA_2 are 9.00 and 9.15 mN*m²/g respectively, which are higher than 7.45 mN*m²/g of the 1% CAT.N1. The results indicate that the combination of the medium charge cationic starch CAT.N1 with either PLA_1 or PLA_2 provides benefits for tear strength when the high kappa number pulp is used.

Effect on Air and Water Resistance Properties

Figure 2 presents the effect of starch CAT.N1 to the air and water resistance properties of paper made from the two different pulps, Pulp A and Pulp B.

For Gurley porosity, it can be observed from Figure 2 that the highest value is recorded for Pulp A + 0.9% CAT.N1 + 0.1% PLA_1, at 16.50 seconds. Similarly, Pulp B + 0.9% CAT.N1 + 0.1% PLA_1 also shows high value at 16.00 seconds which is close to the Pulp A. The lowest value of porosity at 7.74 seconds is seen for the combination of Pulp A with CAT.N1 and PLA_2. These results propose that the presence of PLA_1 helped to fill pores in the paper structure, leading to a lower rate of air passing through paper structure.

For water drop results, Pulp A + 0.9% CAT.N1 + 0.1% PLA_1 still shows the longest time at 224.00 seconds. This is very a significant increase compared to paper made with 1% CAT.N1 and the blank. However, when CAT.N1 combines with PLA_2, the time for one drop of water passing paper thickness decreases rapidly to 138.00 seconds. Once again, the results suggest that with the presence of hydrophobic PLA_1 fiber surfaces are covered resulting in decreasing the rate of water penetrated through paper's thickness.

Effect of Starch CAT.N2 on Paper Properties of Pulps A and B

Effect on Strength Properties

Figure 3 presents the effects of CAT.N2 on strength properties of paper made from Pulp A and Pulp B.

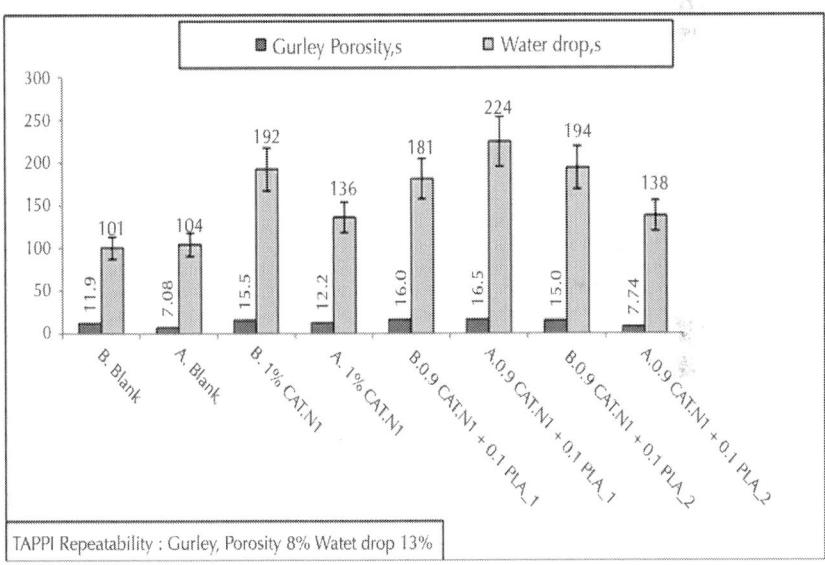

Figure 2: Effects of CAT.N1 on porosity and water drop properties of Pulps A and B.

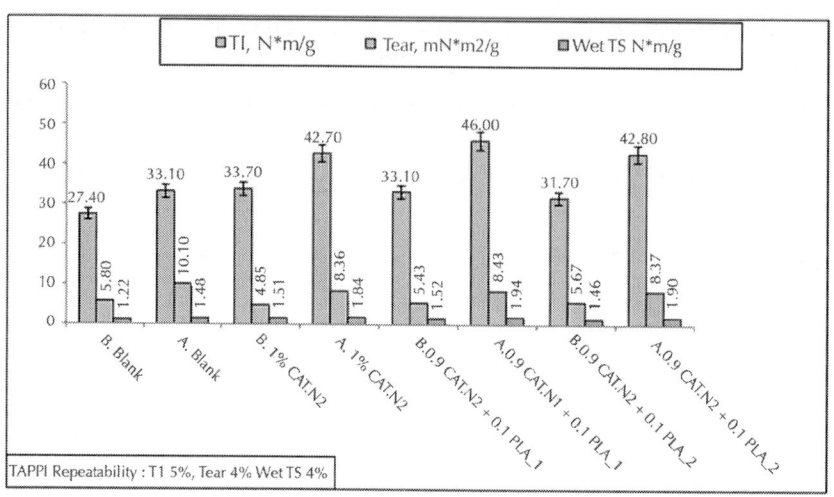

Figure 3: Effects of CAT.N2 on strength properties of Pulps A and B.

With the use of 1% CAT.N2, Pulp A shows the higher increase in TI than Pulp B. When 0.1% PLA_1 is introduced, TI increases from 42.70 to 46.00 Nm/g for Pulp A but slightly decreases from 33.70 to 33.10 Nm/g for Pulp B. However, with the introduction of PLA_2, TI decreases for both paper made from Pulp A and Pulp B. Similarly, wet TS also appear to have the highest value at the combination of Pulp A, CAT.N2 with PLA_1 gives highest TI value.

Unlike the TI and wet TS, with the addition of CAT.N2 and/or PLA_1, PLA_2 tear indices go down for all the cases, including Pulp A and Pulp B compared to the tear of blank handsheet. However, interestingly, for Pulp B, tear strength increases when PLA_1 and PLA_2 are used, compared to the 1% CAT.N2 only. Particularly, tear strength for Pulp A + 0.9 CAT.N2 + 0.1 PLA_1, and Pulp A + 0.9 CAT.N2 + 0.1 PLA_2 are 5.43 and 5.67 mN*m²/g, higher than 4.85 mN*m²/g of the blank. The increase in tear strength is the result of the loss of interfiber bonding strength.

Effect on Air and Water Resistance Properties

Figure 4 presents the effect of CAT.N2 on Gurley porosity and water drop values of paper made from Pulp A and Pulp B.

CAT.N2 gives best response to water resistance ability of Pulp B, in the case of 1% CAT.N2 (231.00 seconds) and 0.9% CAT.N2 + 0.1% PLA_1 (229.00 seconds). Pulp A also shows a rapid increase from 127.00 to 172.10 seconds in water drop value with the use of 0.1% PLA_1. In contrast, the use of PLA_2 with CAT.N2 decreases water drop value for both pulps.

It can also be seen that the highest value of Gurley porosity is shown for the combination of Pulp B + 0.9% CAT.N2 + 0.1% PLA_1. That means the use of pulp with PLA_1 lowers the rate of air passing through the paper structure.

Effects of AS on Paper Properties of Pulps A and B

Effect on Strength Properties

Figure 5 shows the effect of AS on strength properties of paper made from Pulp A and Pulp B.

At 1% AS, pulp A has better response in TI than Pulp B since the increase for Pulp A is ~20% (from 33.10 to 39.70 Nm/g), while the increase for Pulp B is ~9.5% (27.40 to 30.00 Nm/g). When PLA_1 is added, the combination of PLA_1 with AS gives better response to Pulp A than Pulp B. That is because TI increases for Pulp A (to 44.90 Nm/g) but almost keeps unchanged for Pulp B (30.80 Nm/g). Wet TS also give better response for Pulp A than Pulp B with the use of 1% AS and 0.9% AS + 0.1% PLA_1. Tear values get lower than the blank handsheet when AS and PLA are used. That is the result of the increasing fibers bonding strength with the presence of starch and PLA.

Effect on Air and Water Resistance Properties

Figure 6 presents the effect of AS on porosity and water drop values of paper handsheets made from Pulp A and Pulp B.

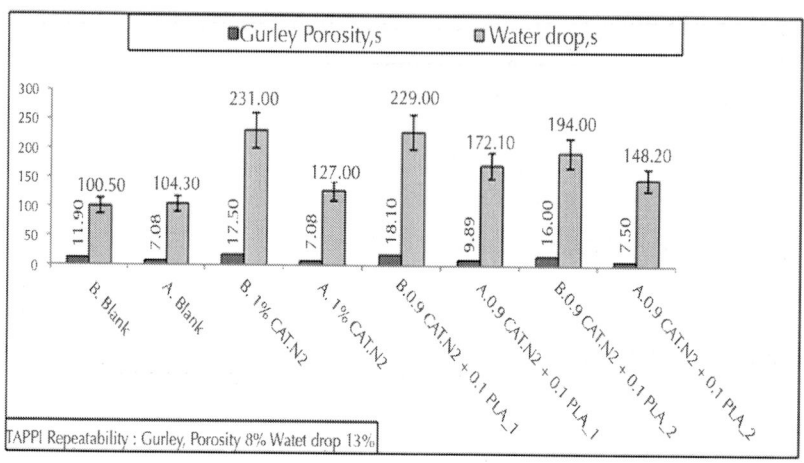

Figure 4: Effects of CAT.N2 on porosity and water drop of Pulps A and B.

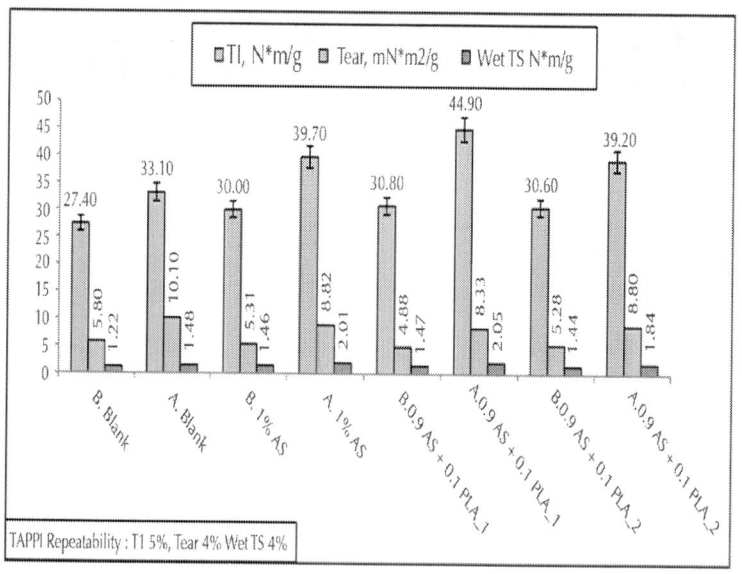

Figure 5: Effects of AS on strength properties of Pulps A and B.

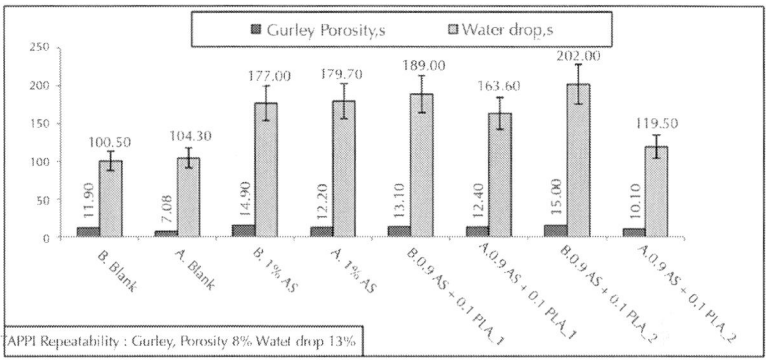

Figure 6: Effect of AS on porosity and water drop of Pulps A and B.

It is clear that at 1% AS water drop value is almost the same at 177.00 and 179.70 seconds for both paper made from Pulp A and Pulp B. When PLA_1 and PLA_2 are introduced at 0.1 %, paper handsheets from Pulp B have a better response than Pulp A. Particularly, the water drop values increase to 189.00 (PLA_1) and to 202.00 (PLA_2) seconds for Pulp B, but decrease to 163.60 (PLA_1) and 119.50 seconds (PLA_2) for Pulp A.

The 1% Gurley porosity also shows higher values for both paper made from Pulp A and Pulp B compared to that of the blank handsheets. The use of PLA_1combining with AS gives a better response for Pulp A, which slightly increases from 12.20 to 12.40 seconds, than for Pulp B which decreases from 14.90 to 13.10 seconds. In contrast to the use of PLA_1, the employment of PLA_2 has the opposite changes. That means Gurley porosity of paper from Pulp B is higher than that of Pulp A.

Effects of Different Starches on Paper Properties of Pulp A

Effect on TI, Tear, Wet TS

Figure 7 shows the effect of different starches (CAT.N1, CAT.N2, AS, ANS) on TI, tear, and wet tensile strength (wet TS) of paper handsheets made from Pulp A.

For the TI, in terms of paper handsheet made with 1% starch, Figure 7 shows that paper handsheets made with CAT.N1 at 1% has highest value of TI at 45.60 Nm/g. The lowest value of TI is recorded for paper handsheet made with 1% ANS (anionic starch). The reason for the different effect is that cationic starch CAT.N1 contains cationic group which increases retention of starch on fibers. That leads to the better bonding degree and hence increasing TI. On the other hand, retention of anionic starch on fibers is less than that of cationic starch. Hence, the fibers bonding degree of paper made from anionic starch and the consequent TI is low.

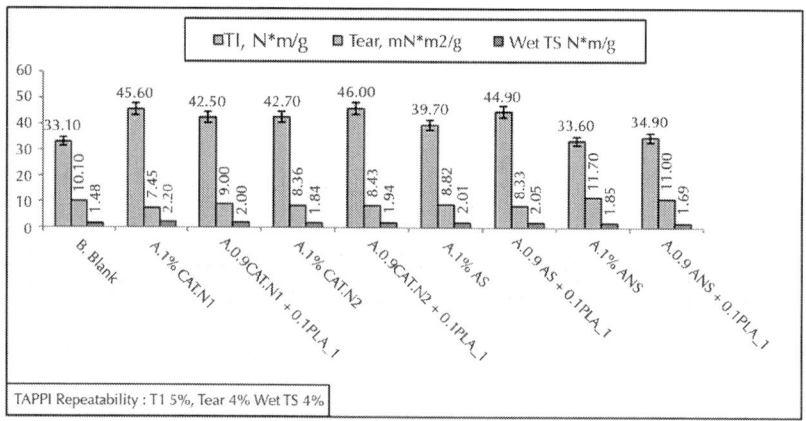

Figure 7: Effects of different starches on strength properties of Pulp A.

Interestingly when combining with 0.1% PLA_1, the TI of paper made with 0.9% CAT.N1 becomes lower than TI of paper made with 1% CAT.N1. In contrast, when combining with 0.1% PLA_1, it can be seen that TIs of paper made with 0.9% CAT.N2 and with 0.9% AS are higher than that of paper made with 1% CAT.N2 and 1% AS, respectively.

For wet TS, the trend is very similar to TIs. It also shows that without PLA_1, paper made with 1% CAT.N1 has highest value of wet TS. However, when combining with PLA_1, wet TS of paper made with CAT.N2 shows the best response.

For tear, for paper made with 1% starch alone the highest tear index is recorded for anionic starch 1% ANS at 11.70 mN*m²/g and the lowest tear index 7.45 mN*m²/g is seen for paper made with 1% CAT.N1. That could be explained that when bonding between fibers increases, fibers become less flexible or more stiff and easy to be torn. When combining with 0.1 % PLA_1, the highest tear index is still recorded for the anionic starch ANS at 11.00 mN*m²/g.

Effect on Air and Water Resistance

Figure 8 shows the effect of four different starches (CAT.N1, CAT.N2, AS, ANS) on air and water resistance of paper handsheets made from Pulp A.

In terms of air resistance, when comparing the four types of starch alone at 1%, both paper handsheets made with amphoteric starch AS and cationic starch CAT.N1 show the longest time at 12.20 seconds for air passing through paper. The shortest time (7.08 seconds) for air to pass through the paper sample is recorded for paper made with CAT. N2.

When combining with 0.1% PLA_1, all the starch-PLA paper handsheets show the increase in time to let air pass through the paper. And the longest time is still recorded for paper made with CAT.N1 (0.9% + 0.1% PLA) at 16.50 seconds. That suggests that with the use of CAT.N1, more pores are filled in the paper structure.

For the water drop result, it can be seen clearly that when combining with 0.1% of PLA_1, cationic starch CAT.N1 also shows the longest time (224.00 seconds) to let one drop of water completely penetrate through the thickness of the paper sample. This suggests that with the present of hydrophobic characteristic of PLA, PLA helps to cover fibers and then repel water or reduce the rate of water penetration into paper.

Effects of Different Starch Contents and Starch-PLAs on TI (Pulp A. CAT.N2)

Figure 9 shows the TI of paper handsheets made from different starches ratios, and from different combination of PLAs with starches.

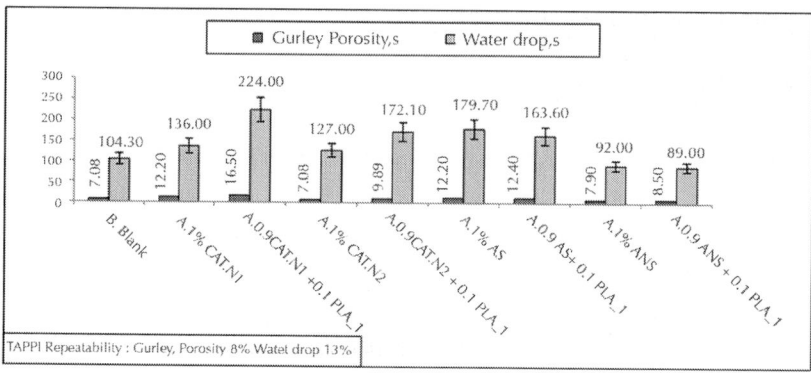

Figure 8: Effects of different starches on porosity and water drop of Pulp A.

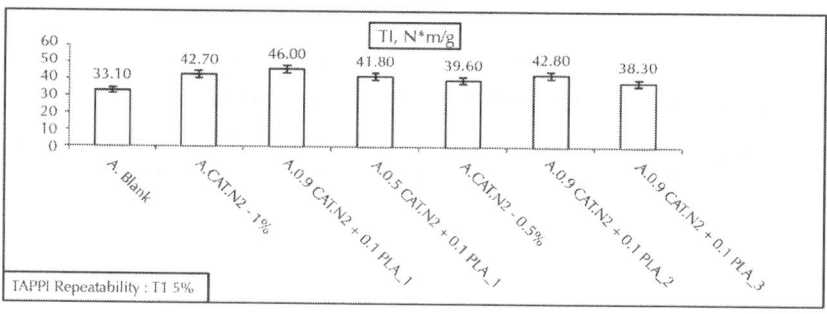

Figure 9: Effects of different starch ratios and starch-PLA on TI of handsheets.

It can be seen from Figure 9 that the highest TI was recorded for the combination (Pulp A + 0.9% CAT.N2 + 0.1% PLA_1) at 46.00 Nm/g. It is greater than the paper made from Pulp A plus 1% CAT.N2 by an amount of 3.3 Nm/g (~7.4% difference). The result suggests that by the substitution of 0.1% PLA_1, the bonding degree of fibers increases which leads to the increase in tensile strength of paper.

It is also clear that with the decrease in starch consumption, the TIs slightly decrease. Particularly, TI drops to 41.80 Nm/g and 39.60 Nm/g for (0.5% CAT.N2 + 0.1% PLA_1) and (0.5% CAT.N2), respectively. The reason for the decrease in TIs is primarily due to the less bonding reinforcement between fibers since less starch is added to the furnish.

Effects of Different PLAs on Paper Properties (Pulp A, CAT.N2)

Figure 10 displays the effect of different PLA MWs on selected properties of paper handsheets made from pulp A and cationic starch CAT.N2.

For the TI, the use of 0.1% PLA_1 with 0.9% CAT.N2 brings paper handsheets the highest value of TI at 46.00 Nm/g while paper handsheets made with 0.1% PLA_3 and 0.9% CAT.N2 have the lowest TI at 38.30 Nm/g. Significantly, the increase of TI for the 1% CAT.N2 compared to the blank is 9.6 Nm/g (42.70 vs. 33.10 Nm/g, about 29% increase). Furthermore, it is clearly seen that TI for the 0.1% PLA_1 with 0.9% CAT.N2 (46.00 Nm/g) is higher about 5.9% than that of the 1% CAT.N2 (42.70 Nm/g). Being lower than TI of paper made with PLA_1, the TI of paper made with 0.9% starch + 0.1 PLA_2 is 42.80 Nm/g which is almost the same with TI of paper made with 1% CAT. N2. The results suggest that PLA_1 and PLA_2 are potential wet end agents to use with starch for tensile strength reinforcement.

Moving to Wet TS, it is clear that wet TS has a very similar trend with TI discussed above. The combination of CAT.N2 and PLA_1 also gives the highest value of wet TS (1.94 Nm/g) while the lowest value is recorded for paper made with PLA_3 (1.83 Nm/g). Therefore, PLA_1 and PLA_2 are also potential agents for wet tensile strength reinforcement.

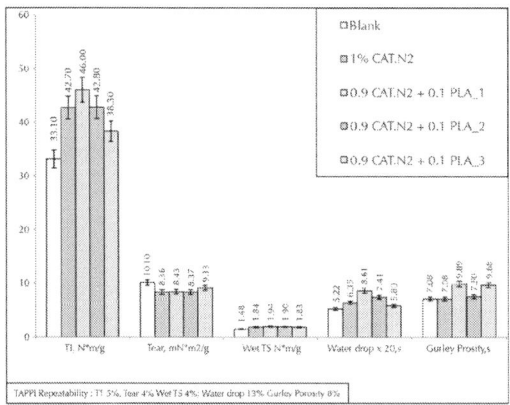

Figure 10: Effects of different PLAs on paper properties (Pulp A, CAT.N2).

For tear result, it can be seen that the highest tear index belongs to the blank experiment. This is because the TI of the blank experiment is lowest, resulting in the increase of tear due to the bonding degree of fibers. Among the tear of paper made from the three PLAs, the use of PLA_3 gives the best value of tear at 9.13 mN*m^2/g. Tear index values of paper made with PLA_1 and PLA_2 are quite similar and are lower than that of paper made with PLA_3.

In regards to water resistance, the longest time for a drop of water to completely penetrate through paper sample is belonged to paper made with PLA_1, at 8.61 × 20 s (172.20 s). The water drop value for paper made with PLA_2 is seen at 7.41 × 20 s (148.20 s), which is higher than the value of 1% CAT.N2 alone. The results also suggest that PLA_1 and PLA_2 are considered to be good agents for water resistance.

Gurley porosity values of paper made from all the three PLAs are clearly higher than that of paper made with no additive and with 1% CAT.N2. Among of the three PLAs, paper handsheet made with PLA_1 and PLA_3 show similar value of Gurley porosity at 9.89 and 9.68 seconds while the lowest value is recorded for paper made with PLA_2 at 7.50 seconds. In general, paper handsheet made from Pulp A with the combination of PLA_1 + CAT.N2 gives the best response for dry tensile strength, wet TS, air and water resistance.

Figure 11 shows a plot of tear and tensile strength of paper handsheet made from Pulp A (high kappa number), with the use of high charge cationic CAT.N2, and the three different molecular weight of PLA. The linear (standard line) shows the normal expectation for tear and tensile relationship.

As can be seen from Figure 11, paper handsheet made from Pulp A + 0.9 CAT.N2 + 0.1 PLA_1 shows the best response of strength, having the highest TI at 46 Nm/g, but also having the tear value of 8.43 MN*m^2/g. Handsheet from Pulp A + CAT.N2 1% only, and handsheet from Pulp A + 0.9% CAT.N2 + 0.1% PLA_2 also have a good correlation of tear and tensile strength, which are seen very close to the line. The blank handsheet shows the highest tear index, but the tensile index is lowest, therefore it is not a good correlation. The results of CAT.N2 and PLA_1 also gives the highest value of wet TS (1.94 Nm/g) while the lowest value is recorded for paper made with PLA_3 (1.83 Nm/g). Therefore, PLA_1 and PLA_2 are also potential agents for wet tensile strength reinforcement.

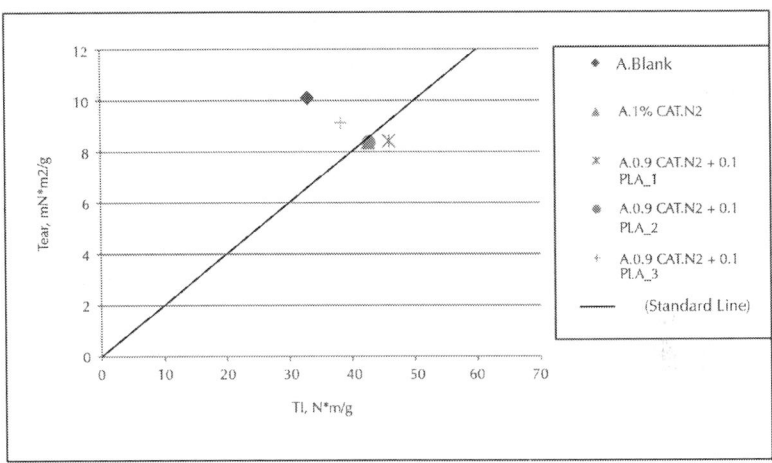

Figure 11: Plot of tear and tensile strength.

Effects of Increasing the Amount of PLA_1 on Tensile Index for Pulp C

Figure 12 shows the effect of different amounts of PLA_1 on TI of Pulp C. Pulp C is the same as Pulp A, but they are refined to different CSFs. Pulp A has CSF of 305 ml while Pulp C has CSF of 298 ml.

It can be seen from Figure 12 that with the use of 0.1% PLA_1 TI of paper made with amphoteric starch AS (45.50 Nm/g) is higher than that (41.00 Nm/g) of paper made with 1% AS. However, with the use of 0.2% of PLA_1, TI of paper made with AS significantly decreases back to 40.70 Nm/g.

For the cationic starch CAT.N1, with the use of 0.1% PLA_1 TI (40.40 Nm/g) is lower than that (44.10 Nm/g) of paper handsheets made with 1% CAT.N1. When 0.2% of PLA_1 is used, TI increases slightly to 42.00 Nm/g but this increase is not significant due to being less than 5% repeatability.

Effect of PLA Used as Surface Sizing Agent on TI and Cobb Value of Linerboard

Figure13 expresses the effect of PLA on TI and Cobb value of linerboard when PLA is used as a surface sizing agent.

As can be seen TI of the paper treated with the highest MW 140,000 - 160,000 is recorded to be greatest at 54.00 Nm/g. Comparing to the blank samples, there is a slight improvement in TI for samples treated with PLA_2 (53.00 vs. 50.30) and PLA_3 (54.00 vs. 50.30).

Linerboard treated with PLA_1 has a lower Cobb value (22.10 g of water/m^2) than the blank sample (23.00 g of water/m^2) which was not treated with PLA. That means PLA_1 has a minor effect in helping paper to resist water. On the other hand, interestingly Cobb values of paper treated with PLA_2 and PLA_3 are seen to have the higher amount of water in the paper at 28.60 and 26.20 g of water/m^2 than the blank at 23.00 g of water/m^2. That suggests that with the present of PLA_2 and PLA_3, the water resistance of paper decreased. One of the possible reasons for the decrease is that PLAs may react with an internal sizing agent AKD, resulting in the reduction in sizing efficiency of AKD for linerboard.

Comparative Effect of the PLAs as Surface Sizing & Wet End Agent on TI

In this study, PLAs were used for two different purposes, as surface sizing agent and wet end agent.Figure 14 illustrates the different effects of PLAs on TI when PLAs plays the different roles.

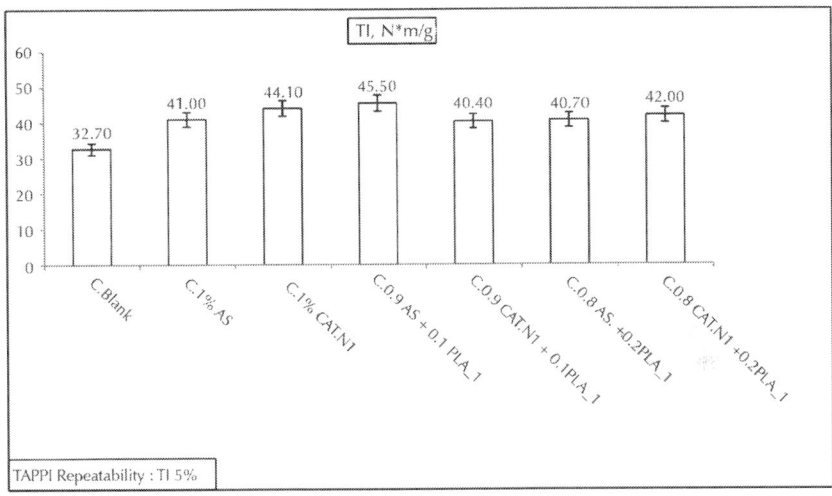

Figure 12: Effects of increasing the amount of PLA_1 on tensile index of Pulp C.

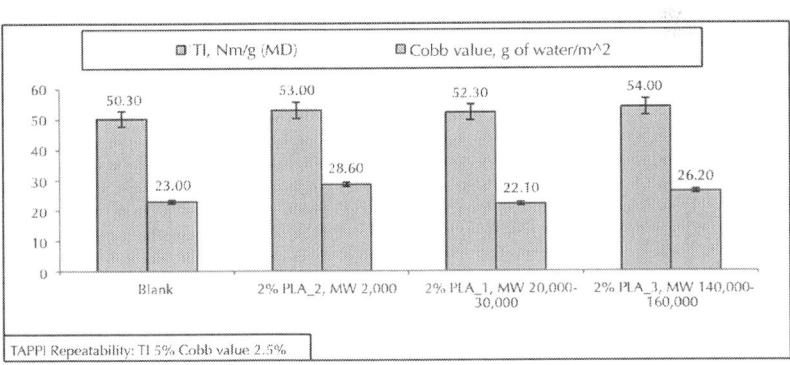

Figure 13: Effect on TI and Cobb value when PLA is used as a surface sizing agent.

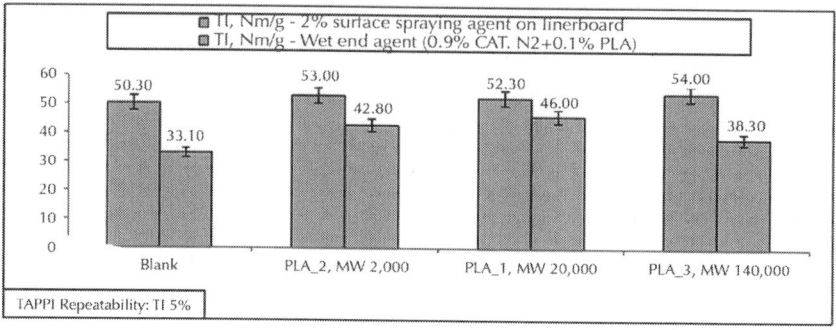

Figure 14: Effect of different PLAs to TI as PLAs are a surface sizing and a wet end agent.

As wet end agent, the graph shows that PLA_1 with MW of 20,000 gives the greatest result in TI at 46.00 Nm/g. Interestingly, the TI (42.80 Nm/g) of paper made from PLA_2 with lowest MW is higher than that (38.30 Nm/g) of paper made from PLA_3 with highest MW (140,000).

As a surface sizing agent on linerboard, the graph shows slight increase in TIs for all the three different PLAs compared to TI of blank which was not treated with PLA. However, it can be seen that there is no significant difference in TI when comparing the three PLAs with each other. In other words, the difference in TI of each pair is less than 5% repeatability. Therefore, it can be concluded that PLAs are not important in the case of a surface sizing agent for linerboard.

CONCLUSIONS

Our results indicate that the effect of PLAs on the selected properties including strength of paper handsheets made from the three different pulps depends on the type of starch, on the molecular weight of PLA, and on the lignin content in the pulps.

- The paper handsheets made from higher kappa number pulp (Pulp A), combining with the medium molecular weight PLA at 0.1% (PLA_1, 20,000 - 30,000), and with higher charge cationic starch at 0.9% (CAT.N2, %N = 0.43) gives the highest improvement for tensile strength, wet tensile strength, air and water resistance. This result verifies that higher kappa number pulp has better attraction to the hydrophobic PLA. Moreover, the results also show that the higher charge cationic starch is retained more on fibers and associated more with negative surface charge of fibers.

- For paper handsheets made from the higher kappa number pulp, with the use of 0.1% PLA_1 the values of Zeta Potential and pH show that the white water suspension is stable and is similar to the values in the condition that no PLA_1 is used. Another conclusion is that the use of 0.1% PLA_1 + 0.9% CAT.N1 is better than the use of 0.2% PLA_1 + 0.8% CAT.N1 for tensile strength.

- When PLAs are used as surface sizing agents for linerboard, the improvement in tensile strength for paper treated with PLA_2 (MW 2000) and PLA_3 (MW 140,000) is higher than that of paper treated with PLA_1. On the other hand, when PLAs are used as wet end agents, the PLA_1 gives better response than PLA_2 and PLA_3 in terms of tensile strength.

- The result of this study interestingly shows that when lignin-rich pulp is employed, tensile strength of paper handsheet made with the combination of amphoteric starch + PLA_1 is very comparable with that of paper handsheet made with high cationic charge starch + PLA_1. This result verifies that, amphoteric starch is a very promising substitution for high cationic charge starch when combining with the medium molecule weight PLA to improve tensile strength of paper.

ACKNOWLEDGEMENTS

The authors are grateful for the support of the Department of Paper and Bioprocess Engineering at the State University of New York, College of Environmental Science and Forestry.

REFERENCES

1. Scott, W.E. (1996) Principles of Wet End Chemistry. TAPPI Press.

2. Biermann, C.J. (1996) Handbook of Pulping and Papermaking. 2nd Edition, Academic Press Limited, London.

3. Neimo, L. (1999) Papermaking Science and Technology: Papermaking Chemistry. Fapet Oy, Helsinki, 268-301.

4.   Gong, C., Hasan, A., Bujanovic, B.M. and Amidon, T.E. (2012) Novel Blend of Biorenewable Wet-End Paper Agents. APPI, 11, 41-48.

5.   Hasan, A., Bujanovic, B.M. and Amidon, T.E. (2010) Strength Properties of Kraft Pulp Produced from Hot-Water Extracted Wood Chips within the Biorefinery. Journal of Biobased Materials and Bioenergy, 4, 46-52. http://dx.doi.org/10.1166/jbmb.2010.1064

6.   Smook, G.A. (2002) Handbook for Pulp and Paper Technologists. 3rd Edition, Angus Wilde Publications, Inc., 207.

Defects Interaction on the Mechanical Properties during Transition Formation of (Mo, Cr) 3 Si Intermetallic Alloys

I. Rosales[1] and H. Martínez[2]

[1]Centro de Investigación en Ingeniería y Ciencias Aplicadas-FCQ e Ing UAEM, Av. Universidad 1001, Col. Chamilpa, Cuernavaca, México

[2]Instituto de Ciencias Físicas, Universidad Nacional Autónoma de México, Cuernavaca, México

ABSTRACT

Molybdenum silicides alloys with different Mo and Cr additions were produced by the arc cast method. The microstructure revealed mostly single phase structure. Mechanical properties were evaluated in the alloys, showing a decreasing behavior on microhardness. Fracture toughness values were obtained from cracks produced by Vickers indentation technique, showing that ternary alloying did not have a significant effect. Vacancy studies demonstrated that thermal vacancies along the transition line slightly affected the mechanical behavior.

INTRODUCTION

Molybdenum silicides have received considerable attention for several authors, in such way that these compounds are interesting due to their good response under severe conditions; these authors evaluated the alloys principally under high temperature for applications as furnace elements [1] - [5]. On the other hand, good mechanical properties are also necessary in these materials. Shah et al. [6] investigated the intermetallics Nb_3Al and Cr_3Si, evaluated their mechanical properties at room and high temperature and reported that Cr_3Si presented an excellent behavior at high temperature, but with limited possibilities of being alloyed. Raj [7] [8] evaluated the mechanical properties of Cr_3Si and $Cr-Cr_3Si$ produced by arc cast and powder metallurgy, where the room temperature fracture toughness was determined in the range of 2.0 and 3.0 MPa $m^{1/2}$, which was comparable with the value reported for Mo_3Si [9] . Fleischer et al. [10] studied the mechanical properties of similar intermetallic compounds. They reported the microhardness value of 1200 HVN at room temperature for Cr_3Si, which was lower than the value reported for Mo_3Si [9]. The analysis of the transition of Mo_3Si to Cr_3Si may provide important information about the optimal concentration of Cr to improve the ductility of the Mo_3Si considering the thermal defects affectation. At present, the mechanical and physical properties of this dual phase intermetallic compound have not been fully explored and therefore limited information is available. The purpose of this work is to find the composition range, in which the best alloy performance is produced that may reduce the Mo_3Si brittleness, correlating the thermal defects affectation on the mechanical properties.

EXPERIMENTAL PROCEDURES

Alloys with Silicon, Molybdenum and Chromium concentrations, were prepared by arc-melting of nominally pure elements in a partial pressure of argon (99.99% purity). The alloys were drop-cast into water-cooled copper rod molds with a diameter of 6.5 mm and 10mm length. Table 1 shows the nominal compositions for Si and Cr, obtained by assuming that the weight loss during melting were due to evaporation of Si and Cr, so an extra 0.5 at.% of this elements were added to compensate this weight losses. The specimens were annealed in a vacuum of 10^{-4} Pa for 24 h at 1400°C, and then furnace cooled using a cooling rate of 2.5°C·s⁻¹ between 1400°C and 1000°C. After metallographic polishing, the specimens were etched with Murakami's reagent for 1 - 2 s. The etched specimens were observed in an optical microscope. The lattice parameters were determined by X-ray diffraction of powders with a size < 45 **μm** and internal silicon standard. The experimental accuracy of the lattice parameter measurements was estimated to be 3×10^{-4} **Å**. Density values were calculated from the lattice parameter obtained by cell refinement of each alloy and bulk densities were obtained by pycnometric measurements. Vacancy concentrations were calculated from the follow expression [11]:

$$C_v = \frac{\left(\rho_{\text{X-ray}} - \rho_B\right)}{\rho_B}$$

Where: C_v = Vacancy concentration;

$r_{\text{X-ray}}$ = Density calculated by X-ray diffraction analysis;

r_B = Bulk Density obtained by He Picnometric measurements.

The microhardness was measured on a Buehler microhardness tester using a load of 500 g with a holding time of 15 s. Fracture toughness evaluation was performed using the indentation method with a Vicker's indenter [11].

RESULTS AND DISCUSSION [9]

Microstructures and X-Ray Diffraction

The microstructures of the samples with different Cr content are shown in Figures 1(a)-(d). The single-phase microstructure can be seen along the different ternary alloy compositions, grain boundaries are on the order of 120 **µm** averages. Surfaces crack free were obtained by this method in all range of compositions (dark dots observed on the surface sample probably are impurities from the crucible). Figure 2 shows the X-ray diffraction patterns obtained from the samples with different Cr compositions. The observed silicon peaks correspond to the standard reference material used to determine the lattice parameter. The diffraction pattern shows a clear transition of the Mo_3Si to Cr_3Si, at 2θ = 46 degrees and after 36 Cr at.%; furthermore, it is observed that the peaks show an increased Bragg angle respect to the sample with 17 Cr at.%, which imply a lattice parameter reduction, no additional phase was observed, except for Mo (211) at 2θ = 73 degrees (see spectrum 2 in Figure 2), therefore, it is concluded that the main constitution phases along the transition range are the prototypes Cr_3Si and Mo_3Si which possesses the A15 structure which coexist both in the range of 36 and 60 Cr at.%. These phases are in good agreement with the Cr-Mo-Si ternary phase diagram.

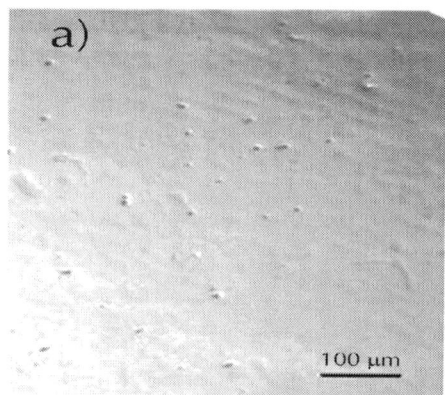

(a)

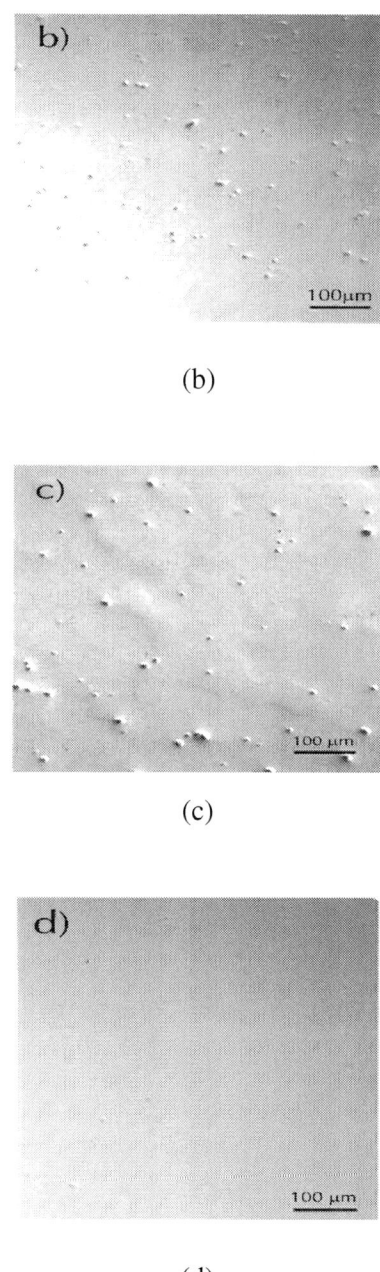

Figure 1: Micrographs of the Cr alloys with a) 17 at. %, b) 36 at. %, c) 60 at. % and d) 76 at. %.

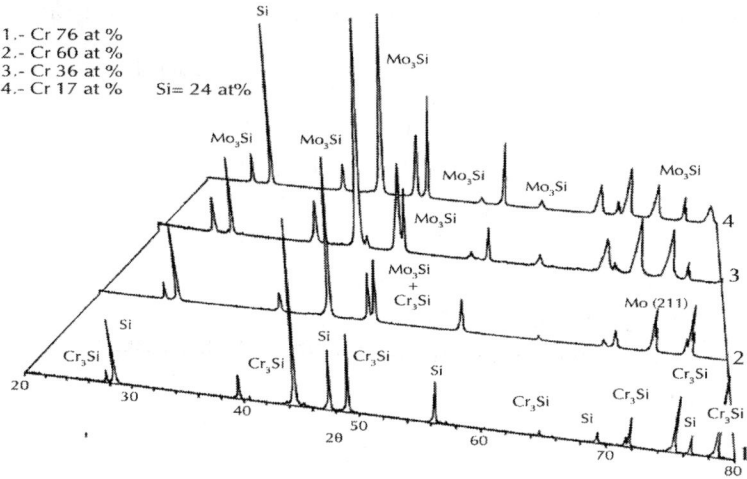

Figure 2: X-ray diffraction pattern from samples with Cr content.

Mechanical Properties

Hardness Evaluation

Figure 3 shows the plots of the values obtained at room temperature for the microhardness as a function of the Cr concentration. For the alloys with Cr content, the microhardness obtained, presents a decreasing behavior as the Cr at. % increases. In spite of the difference in size of the added atom with respect to the atomic radii of Mo (1.37 Å), it is deduced that Cr (1.25 Å) atoms remain in substitution places in the unit cell, affecting the hardness, due to the lattice distortion that can produce an arrest mode in the dislocation mobility. Fleischer [10] reports a value of 1200 HV at room temperature in samples of Cr_3Si, which is lower (less than 10%) than the results obtained in this study. The hardness obtained in this alloy suggests that one potential application could be in component parts exposed to friction conditions, perhaps by using powder depositions techniques. High hardening can be attributed to an excess of vacancies [12], however, this is not the case for our study, because the vacancy concentration is minimum.

Fracture Toughness Evaluation

Figure 4 presents the values of fracture toughness against the different Cr concentrations. For the alloys with Cr additions the value of fracture toughness decreases as a function of Cr concentration. A reduction of about 40% is observed in the fracture toughness between 16 and 76 Cr at. %. This reduction is approximately of 35% in comparison with the value reported for Mo_3Si single phase [9]. It is well known that thermal vacancies may induce some grade of softening in B2 intermetallic compounds [13] and this phenomenon is referred to a solid solution addition. In our case, we have an A15 structure which contain an arrangement of six Mo atoms plus two Si atoms for the Mo_3Si (similar array with the Cr_3Si) which produces a complex structure, thus, the decreasing behavior of the fracture toughness as a function of the Cr concentration until 48 Cr at. % content indicates that the presence of point defects can be quite substantial and therefore an effect in the fracture toughness of the materials attributed to this point defect may happen. Other complementary explanation about the toughness increment for sample S4, may be attributed to the interaction of Mo atoms with Cr atoms which are distributed along the A15 lattice where coexists both phases as is shown in the X-ray analyses. Also it has to be mentioned that, since both phases possesses A15 structure, this fact can be considered as highly positive due to the coherency of the lattice parameter, that definitely contribute to reach some grade of plasticity of the alloy.

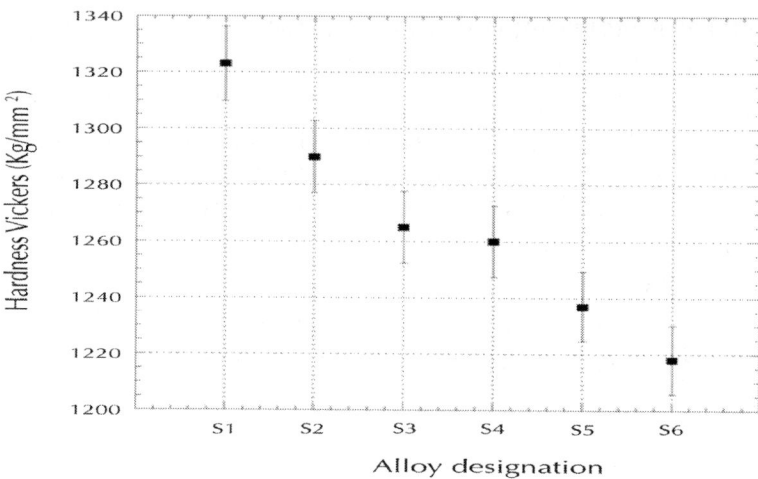

Figure 3: Microhardness as a function of Cr concentration.

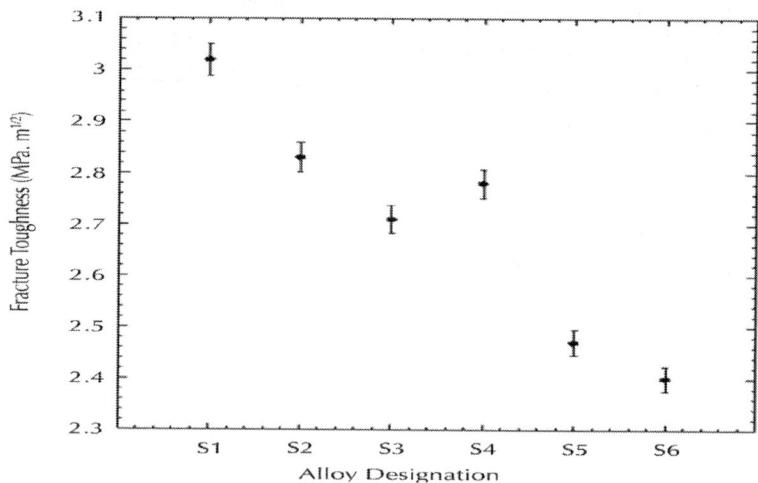

Figure 4: Fracture toughness against different Cr concentrations.

Defects Analysis

In Figure 5, we present the results of the lattice parameter data as a function of the Cr content. The solid line represents a second order polynomial fit through the lattice parameter data. According to a Rydberg study [14], it is possible to correlate hardness with the reciprocal size of the atoms for pure elements. This conclusion and the underlying experimental results give, in principle, the explanation that the hardness is proportional to the cohesive forces. In the present work, the hardness has a decreasing behavior as a function of the Cr concentration and the lattices parameter decreases with the Cr concentration, so then the relation between hardness and lattice parameter is not reciprocal. Therefore, the present results cannot be explained by this argument. From Table 1 it can be seen that the density measurements of the alloys with Cr additions have a decreasing behavior as a function of the Cr content.

The Goldschmidt radius of Mo is significantly larger than the Cr radii [15]. Consequently, as seen in Figure 5, the Cr_3Si lattice parameter decreases significantly.

The lattice mismatch between Mo_3Si and Cr_3Si of 6.94% almost corresponds to the change of the radius of Mo to Cr (8.76%). On the other hand, in order to obtain a preliminary understanding of the trend of the lattice parameter as a function of the Cr content in the alloy of Mo_3Si, one recalls that the lattice parameter of substitutional solid solutions is usually an average of the interatomic spacing in the pure components weighted according to the atomic fractions present; this observation is known as Vergard's law [16]. We could start with this law to estimate the lattice parameter as a function of the at. %. of Cr in the Mo_3Si alloy The estimation using this law was carried out in the following steps:

- The values of the lattice parameter of the alloys Mo_3Si, and Cr_3Si, were taken from the experimental results;
- The lattice parameter as a function of the Cr content was evaluated from Vergard's law, using the values of the lattice parameter of the alloys Mo_3Si and Cr_3Si.

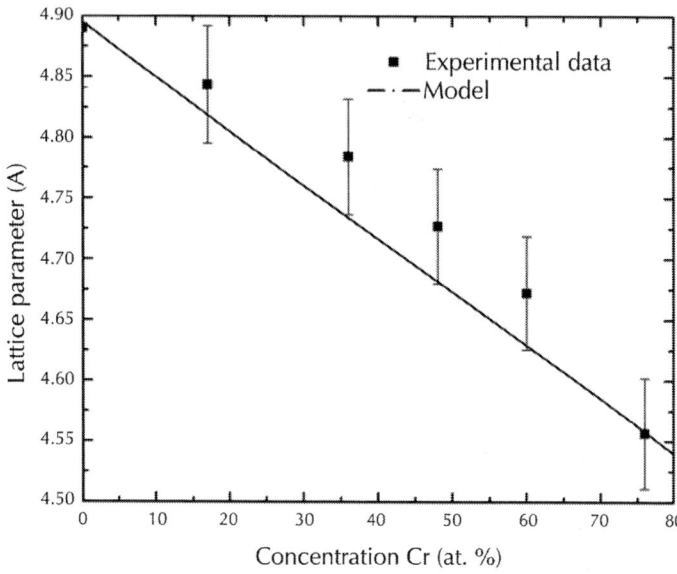

Figure 5: Lattice parameter as a function of Cr, experimentally and using the proposed model.

Table 1: Alloy composition, lattice parameter and densities for the alloys as a function of Mo and Cr additions.

Alloy Designation	Si [at. %]	Mo [at. %]	Cr [at. %]	Lattice Parameter [Å]	X-Ray Density [g/cm3]	Bulk Density [g/cm3]	Cv [%]
S1	24.00	76.00	0.00	4.8900	9.00035	8.9985	0.020
S2	24.00	59.00	17.00	4.8436	8.39891	8.3957	0.038
S3	24.00	40.00	36.00	4.7842	7.71521	7.7123	0.037
S4	24.00	28.00	48.00	4.7270	7.33312	7.3302	0.039
S5	24.00	16.00	60.00	4.6721	6.92713	6.9254	0.024
S6	24.00	0.00	76.00	4.5567	6.49180	6.4873	0.069

Although the estimation has some limitations, the model could be used as a first approximation. The results of these estimations are shown in Figure 5. Here it can be seen that the estimations and the present measurements are in good agreement. These results permit us to infer that the process for the case of Cr is clearly substitutional.

Table 1 shows the lattice parameter and densities values obtained from X-ray diffraction data analysis for several Cr content, using silicon peaks as references. In the case of alloys with Cr additions, the density results shown a reduction when the Mo content is substituted for the Cr addition (which has a lower atomic mass weight) until the stoichiometric composition of Cr with Si is reached. From the vacancy concentration analysis it is concluded that do not exist enough amount of this defects to promote the lattice deformation, which can be detrimental in the mechanical properties obtained.

CONCLUSIONS

Ternary alloys with Cr additions on Mo_3Si matrix were produced. The hardness of the alloys in the present study falls in the range of 1200 - 1330 Kg/mm^2. Fracture toughness of the alloys with different Cr additions increases as much as 25%. Density values indicate a low level of thermal defects existing in the alloys. From the studies of the lattice parameter it can be deduced that the variation obtained is attributed to the distortion of the lattice when an atom with different atomic radii is added, not for vacancies concentrations, performing a substitutional process for the case of Cr and for consequence affecting in positive way the mechanical properties during the transition process in a specific range Cr concentration.

ACKNOWLEDGEMENTS

We are grateful with R. Guardian for their technical assistance. Also thanks to ORNL for the use of their facilities in sample preparation. This research was partially sponsored by PROMEP PTC-074.

REFERENCES

1.   Vasudevan, A.K. and Petrovic, J.J. (1992) A Comparative Overview of Molybdenum Disilicide Composites. Materials Science and Engineering: A, 155, 1-17. http://dx.doi.org/10.1016/0921-5093(92)90308-N

2. Petrovic, J.J. and Vasudevan, A.K. (1999) Key Developments in High Temperature Structural Silicides. Materials Science and Engineering: A, 261, 1-5.http://dx.doi.org/10.1016/S0921-5093(98)01043-0

3. Akinc, M., Meyer, M.K., Kramer, M.J., Thom, J.A., Huebsch, J.J. and Cook, B. (1999) Boron-Doped Molybdenum Silicides for Structural. Materials Science and Engineering: A, 261, 16-23. http://dx.doi.org/10.1016/S0921-5093(98)01045-4

4. Schneibel, J.H., Kramer, M.J. and Easton. D.S. (2002) A Mo-Si-B Intermetallic Alloy with a Continuous α-Mo Matrix. Scripta Materialia, 46, 217-221.http://dx.doi.org/10.1016/S1359-6462(01)01227-1

5. Liu, C.T., Schneibel, J.H. and Heatherly, L. (1999) High Temperature Ordered Intermetallic Alloys VIII. Materials Research Society Symposium Proceedings, 552.

6. Sha, D.M. and Anton, D.L. (1992) Evaluation of Refractory Intermetallics with A15 Structure for High Temperature Structural Applications. Materials Science and Engineering: A, 153, 402-409. http://dx.doi.org/10.1016/0921-5093(92)90228-S

7. Raj, S.V., Whittenberg, J.D., Zeumer, B. and Sauthoff, G. (1999) Elevated Temperature Deformation of Cr_3Si Alloyed with Mo. Intermetallics, 7, 743-755.http://dx.doi.org/10.1016/S0966-9795(98)00095-8

8. Raj, S.V. (1995) An Evaluation of the Properties of Cr_3Si Alloyed with Mo. Materials Science and Engineering: A, 201, 229-241. http://dx.doi.org/10.1016/0921-5093(95)09767-8

9. Rosales, I. and Schneibel, J.H. (2000) Stoichiometry and Mechanical Properties of Mo_3Si. Intermetallics, 8, 885-889.

10. Fleischer, R.L. and Zabala R.J. (1989) Report No. 89CRD201. General Electric Research & Development Center, Schenectady.

11. Zhu, J.H., Pike, L.M., Liu, C.T. and Liaw, P.K. (1999) Point Defects in Binary Laves Phase Alloys. Acta Materialia, 47, 2003-2018. http://dx.doi.org/10.1016/S1359-6454(99)00090-7

12. Jordan, J.L. and Deevi, S.C. (2003) Vacancy Formation and Effects in FeAl. Intermetallics, 11, 507-528.

13. Pike, L.M., Chang, Y.A. and Liu, C.I. (1997) Point Defect Concentrations and Hardening in Binary B2 Intermetallics. Acta

Materialia, 45, 3709-3719.http://dx.doi.org/10.1016/S1359-6454(97)00028-1

14. O'Neill, H. (1967) Hardness Measurement of Metals and Alloys. 2nd Edition, Chapman and Hall, London.

15. Laves, F. (1956) Theory of Alloy Phases. American Society for Metals, Philadelphia, 131-133.

16. Ohring, M. (1995) Engineering Materials Science. Academic Press, Waltham.

10

Fluid Filtration and Rheological Properties of Nanoparticle Additive and Intercalated Clay Hybrid Bentonite Drilling Fluids

Matthew M. Barry[a], Youngsoo Jung[a], Jung-Kun Lee[a], Tran X. Phuoc[a, b], and Minking K. Chyu[a]

[a]Department of Mechanical Engineering and Materials Science, University of Pittsburgh, Pittsburgh, PA 15261, USA

[b]National Energy Technology Laboratory, U.S. Department of Energy, Pittsburgh, PA 15236, USA

ABSTRACT

The fluid filtration and rheological properties of low solid content (LSC) bentonite fluids containing iron-oxide (Fe_2O_3) nanoparticle (NP) additives and two different NP intercalated clay hybrids, iron-oxide clay hybrid (ICH) and aluminosilicate clay hybrid (ASCH), under both low-temperature low-pressure (LTLP: 25 °C, 6.9 bar) and high-temperature high-pressure (HTHP: 200 °C, 70 bar) conditions are investigated. The viscosity of each fluid was measured under LTLP and HTHP conditions using a pressurized and heated rotational viscometer. The LTLP and HTHP fluid filtrate volumes were measured in accordance to American Petroleum Institute standards. The addition of ICH and ASCH into bentonite solutions reduced both LTLP and HTHP fluid loss as much as 37% and 47% as compared to the control, under the respective conditions. The pure addition of 0.5 wt% 3 and 30 nm Fe_2O_3 NP increased the LTLP fluid filtration as much as 14% as compared to the control. However, this addition of Fe_2O_3 NP decreased the HTHP fluid filtrate volumes as much as 28% as compared to the control. It was found that the addition of clay hybrids reduced LTLP and HTHP fluid loss due to a restructured mode of clay platelet interaction attributed to a modification in surface charge as demonstrated by zeta potential measurements and scanning electron microscope images.

INTRODUCTION

Bentonite is a montomorillonite (Mt) clay commonly used in conventional water-based drilling fluids due to inherent, well-performing rheological properties and the vast knowledge thereof (Bourgoyne et al., 1986 and Luckham and Rossi, 1999). Mt clay solutions exhibit shear-thinning viscosities which provide good pumpability and carrying capacity of cuttings (Bourgoyne et al., 1986). Additionally, Mt clay forms relatively low permeability filter cakes on the borehole surfaces (Larsen, 1955); thick filter cake formations reduce the effective diameter of the borehole, which causes excessive drag, torque losses, high swab and surge pressures and increases the risk of differential sticking of the drill string (Darley and Gray, 1988). Mt clay begins to chemically break down at temperatures as low as 120 °C (Kelessidis et al., 2006), increasing the drilling fluid loss into sub-surface formations and reducing the effective carrying capacity of cuttings (Bourgoyne et al., 1986), making these fluids ineffective for high-temperature and high-pressure (HTHP) drilling applications. High-temperature and high-pressure applications can be classified into three categories, defined by bottom hole temperature (BHT) and pressure (BHP); HTHP when BHT is greater than 150 °C and less than 205 °C or when BHP is greater than 690 bar and less than 1380 bar; ultra-HTHP when BHT exceeds 205 °C and is less than 260 °C or when BHP exceeds 1380 bar and is less than 2410 bar; HTHP-hc when BHT is greater than 260 °C or when BHP is greater than 2410 bar (Belani and Orr, 2008).

To extend the temperature limitation of clay-based drilling fluids, lignite is used within water-based Mt solutions to reduce fluid loss at temperatures up to 177 °C (Kelessidis et al., 2007). Above the temperature limitations of lignite-Mt solutions, non-aqueous drilling fluids are preferred in field applications for they are inherently lubricious and provide stable rheological properties up to operating temperatures of 204 °C (Mas et al., 1999). However, it is desired to increase the operating limitations of clay-based drilling fluids for HTHP conditions.

To increase the temperature limitation of clay-based drilling fluids, methods have been proposed to modify surface charge to yield lesser fluid filtration loss under HTHP conditions. Haloing stabilizes colloidal fluids such as gels (Tohver et al., 2001), and this study has been extrapolated to surface cation exchange within drilling fluids to promote particular surface structures (Tombácz et al., 1989 and Tombácz et al., 2001). The use of organic polymers as viscosifiers suggests cost-effective drilling fluid improvements under HTHP conditions, but such organic polymers also have temperature limitations (Mahto and Sharma, 2004).

It has been demonstrated that micro and nanoparticles (NP) can maintain or increase the performance of drilling fluids under HTHP conditions (Baird and Walz, 2007). Two methods of micro and NP inclusion into drilling fluids are infusion and intercalation (Huang et al., 2010 and Jung et al., 2011). Infusion is the mere addition of NP into the solution, without much control over the chemical and mechanical interaction between the NP and clay platelet. Micro and NP additives within drilling fluids have the potential to plug pores of subsurface formations, such as shale, preventing drilling fluid loss into said formations and subsurface fluid intrusion into the borehole without compromising the extraction of fluids from the productive horizon (Srivatsa and Ziaja, 2011). Intercalation is the process of modifying interlaminar cations and neutral species clays, and in this instance, of the Mt aluminosilicates with various alkalies and NP, resulting in permanently charged clay platelets. Intercalation of NP into a host matrix is a widely applied approach for modifying the structure of materials, resulting in improved chemical and physical properties. Inorganic NP have been used to stabilize fluid viscosity at high temperatures via manipulating and modifying the surface morphology and reactivity of the clay platelets (Huang and Crews, 2008). Although much focus has been placed on the rheological properties of intercalated-bentonite solutions (Laribi et al., 2006 and Tombáccz and Szekeres, 2004), current literature does not reflect studies on the fluid filtration losses under LTLP and HTHP conditions.

This work is to address the need of concurrently studying the rheological properties and fluid filtration losses of low-solid content (LSC) bentonite fluids with and without NP additives and embedded clay hybrid solutions under LTLP and HTHP conditions. LSC bentonite fluids with iron-oxide (Fe_2O_3) additives and two different NP embedded clay hybrids are investigated under both low-temperature low-pressure (25 °C, 6.9 bar) and HTHP (200 °C, 70 bar) conditions. The NP embedded clay hybrid solutions of interest are iron-oxide clay hybrid (ICH) and Al_2O_3–SiO_2 clay hybrid (ASCH). It will be shown that added ICH and ASCH particles in the LSC bentonite fluid system decreased filtration volumes as compared to a control bentonite solution under LTLP and HTHP conditions due to decreased filter cake permeability. Permeability of the produced filter cake is determined by the mode of inter-platelet association. Furthermore, it will be demonstrated the pure addition of Fe_2O_3 NP in concentration of 0.5 wt% with sizes of 3 and 30 nm had variable effects on fluid loss.

EXPERIMENTAL SET-UP, DRILLING FLUID PREPARATION, AND TESTING

The experiments focused on LTLP and HTHP nanoparticle drilling fluid performance via filtration testing. Low-temperature low-pressure static filtration fluid loss testing was performed in accordance to American Petroleum Institute (API) standards (Bourgoyne et al., 1986, Recommended Practice Standard Procedure for Laboratory Testing Drilling Fluids, 1990 and Spec 13A, 1993). Figure 1(a) and (b) illustrates both the LTLP and HTHP filtration testing system. Five different configurations of drilling fluid solutions were prepared, each made with a five weight-percent (5.0 wt%) bentonite in deionized water solution, as summarized in Table 1; a control of 5.0 wt% bentonite (Sample A); 5.0 wt% bentonite solution with 0.5 wt% 3 nm Fe_2O_3 NP

(Sample B); 5.0 wt% bentonite solution with 0.5 wt% 30 nm Fe_2O_3 NP (Sample C); 5.0 wt% bentonite solution with 0.5 wt% ICH (Sample D); 5.0 wt% bentonite solution with 0.5 wt% ASCH (Sample E). Table 6 provides conversions from SI to U.S. customary units.

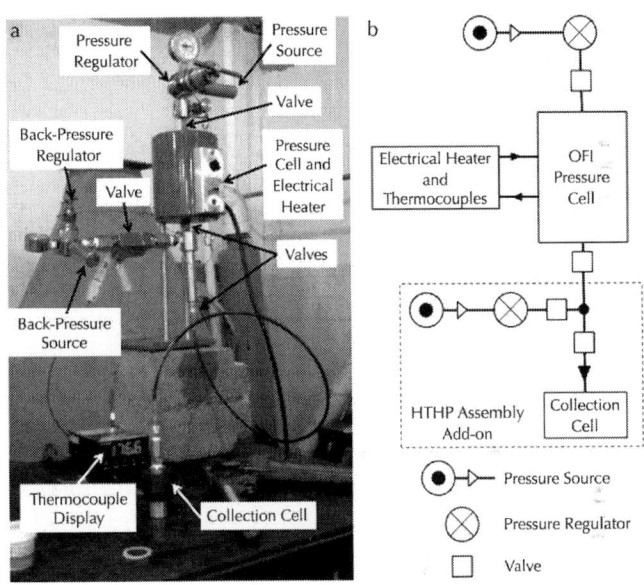

Figure 1: =Low-temperature low-pressure and high-temperature high-pressure filtration testing assemble with (a) HTHP experimental set-up and (b) schematic of LTLP and HTHP equipment.

Table 1: Abbreviated sample names

Sample	Sample abbreviation
5.0 wt% bentonite	Sample A
5.0 wt% bentonite with 0.5 wt% 3 nm Fe_2O_3	Sample B
5.0 wt% bentonite with 0.5 wt% 30 nm Fe_2O_3	Sample C
5.0 wt% bentonite with 0.5 wt% ICH	Sample D
5.0 wt% bentonite with 0.5 wt% ASCH	Sample E

Synthesis of Nanoparticle-Clay Hybrids and Additive Solutions

In this study, two different types of NP-clay hybrid particles were prepared through the intercalation of metal polycations into the interlayer space of the clay and subsequent thermal annealing. The Na$^+$-Mt clay (Kunipia F) used was Na$_{0.35}$K$_{0.01}$Ca$_{0.02}$(Si$_{3.89}$Al$_{0.11}$)(Al$_{1.60}$Mg$_{0.32}$Fe$_{0.08}$)O$_{10}$(OH)$_2$·nH$_2$O. A detailed synthesis process of the hybrid particle fabrication can be found in Jung et al. (2011) and Son et al. (2010). A brief description follows here.

An iron polycation solution was prepared by dissolving 0.2 M FeCl$_3$-6H$_2$O into 0.4 M NaOH at 70 °C. The polycation solution was mechanically mixed with Na$^+$-Mt at the same temperature to yield intercalation. Excessive polycations on the surface of the clay platelets were rinsed with deionized water. The resulting product was dried and annealed at 450 °C in a N$_2$ atmosphere to fully transform intercalated iron polycations into embedded iron oxide NP.

The same methodology was employed for the fabrication of Al$_2$O$_3$–SiO$_2$ NP. ASCH particles were created via obtaining an Al polycation solution from an aqueous solution of 0.2 M AlCl$_3$–6H$_2$O and 0.2 M NaOH. Upon 24 h of mechanical mixing at room temperature, the solution was mixed with Si$_{(OC2H5)4}$ to make an aqueous hydroxy silico-aluminum polycation solution. The hydroxy silico-aluminum polycation solution was mixed with the Na$^+$-Mt solution, yielding intercalatation between the prepared polycations into the interlayer of the Na$^+$-Mt. The hybrid particles were then washed with deionized water, dried and annealed at 400 °C in a N$_2$ atmosphere to fully transform the hydroxy silico-aluminum polycations into embedded aluminosilicate NP.

To form a 5.0 wt% bentonite 0.5 wt% hybrid solution, the hybrid particles were mixed with bentonite (H$_2$Al$_2$O$_6$Si) in deionized water. These solutions were mechanically stirred for 30 min at 25 °C. Afterwards, the solutions were sonicated for 30 min at 25 °C. The additive NP solutions were made by the simple mixing and sonication of the aforementioned weight percentages under ambient conditions.

Rheological Testing

A viscometer equipped with a HTHP cell was used to characterize the rheological properties under different shear rates. The viscometer operated at temperatures between 25 °C and 200 °C and pressures between 6.9 bar and 70 bar, coinciding with the LTLP and HTHP filtration environments. The shear rate varied from 1 to 200 s^{-1} with a step size of 4 s^{-1}. Each step was executed for 10 s. Additionally, the zeta potential measurements of the 3 and 30 nm Fe_2O_3 NP solutions were conducted as to determine charge.

Fluid Filtration Testing

The LTLP static filtration testing was conducted using a 175 mL static filter press with a regulated CO_2 pressurization system and reinforced filter paper. The operating pressure was 6.9 bar and the temperature was atmospheric (25 °C) as to replicate the testing conditions of API fluid-loss testing. The mass of the filtrate was recorded per unit time with a sample rate between five and six measurements per second using a digital balance and a personal computer. The volume was then calculated per the average density of the filtrate collected, which allowed the volume versus time to be determined. The filter cakes were gently rinsed with water to remove any excess fluid and then the thickness was measured using a micrometer.

The HTHP static filtration testing was conducted at 70 bar and 200 °C under a N_2 environment. This system used the same filter press but included a modification to the fluid exit valve in which a two-stage valve and back-pressure regulator were included. The modification allowed for a continual, variable back-pressure to be applied on the pressure cell during collection as to ensure a constant 70 bar pressure differential and to avoid evaporation of the fluid from the solution until it cooled. Adjustments were made to the total LTLP and HTHP filtration times as per the recommendation of Arthur et al. (Arthur and Peden, 1988). Each LTLP and HTHP test was conducted three times, with the average values reported. For HTHP testing, the volume was collected at 5, 15, 30, 60 and 90 min. Filter cake thickness measurements under HTHP conditions were not conducted due to the damage (dehydration and cracking) the cake sustained during cooling of the HTHP cell.

The undisturbed filter cakes of both LTLP and HTHP samples were then freeze-dried at -45 °C before being characterized with a field emission electron microscope. The alignment of the clay platelets was studied using SEM to observe porous matrix morphology, as has been done in other sources (Chenevert and Huycke, 1991, Hartmann et al., 1988, Porte, 1980 and Plank et al., 1991). Images were taken in the direction of fluid flow.

RESULTS AND DISCUSSION

This section presents the rheological properties and filtration volumes of various drilling fluids under both LTLP and HTHP conditions. The performance of these novel drilling fluids is investigated in comparison to control of 5.0 wt% bentonite solution.

Fluid Rheological Properties

Drilling fluids are typically characterized as thixotropic shear-thinning solutions, exhibiting viscosity that decreases with an increase in the shear rate (Livescu, 2012). Additionally, these fluids exhibit a yield stress, or the stress that must be applied to the fluid to initiate flow, which was found by extrapolating the shear stress versus shear rate curve to zero-shear conditions. The yield stress values are reported in Table 2. The shear stress versus shear rate plots for LTLP and HTHP conditions are shown in Fig. 2(a) and (b).

Table 2: The yield stress τy, flow consistency index K and flow behavior index n of all drilling fluids under LTLP conditions treated as Herschel–Bulkley fluids

Sample	$\tau y_{(Pa)}$	K(Pa s)	n
A	1.04	0.022±0.001	0.890±0.001
B	2.43 (1.436±0.046)	0.060±0.007	0.742±0.020
C	2.03 (1.403±0.026)	0.033±0.003	0.817±0.015
D	9.85 (3.165±1.411)	3.041±1.07	0.225±0.040
E	0.25	0.011±0.002	0.856±0.030

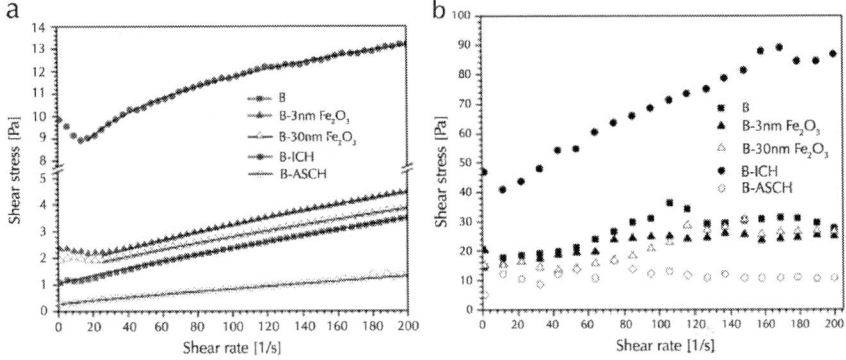

Figure 2: Shear stress versus shear rate of various drilling fluids under (a) LTLP and (b) HTHP conditions. Solid lines indicate Herschel–Bulkley fits, based upon the information provided in Table 2.

Evaluating the drilling fluid as a Herschel–Bulkley fluid, the shear stress τ is

$$\tau = \tau_y + K\dot{\gamma}^n \tag{1}$$

where τ_y is the yield stress, K is the flow consistency index, $\dot{\gamma}$ is the shear rate and n is the flow behavior index. The flow consistency and flow behavior index were found by fitting the shear stress versus shear rate data to a two-equation yield power law (Herschel–Bulkley) with 95% confidence bounds. The flow consistency and behavior indices are presented in Table 2 with the yield stress.

It is noted from Fig. 2(a) that the 3 and 30 nm Fe_2O_3 and ICH solutions exhibit a yield stress, then a decrease in stress with an increase in shear rate until a flow point is reached. This indicates that the fluid appears to be in solid-like elastic state. After the flow point, the stress versus shear rate behaves according to the Herschel–Bulkley model, that is, the fluid has transitioned to a liquid-like viscous state. This is attributed to the complex interparticle interaction of additive Fe_2O_3 NP and ICH particles within the bentonite solution, which is elaborated upon in the following paragraphs. The values for yield stress presented in Table 2 with parentheses indicate the modified yield stress (flow point) for which the fluid was fitted.

The ICH and additive Fe_2O_3 NP solutions have greater viscosities whereas the ASCH has a lower viscosity as compared to the bentonite control under LTLP conditions. Under HTHP conditions, the ICH solution has a greater viscosity whereas the Fe_2O_3 additive and ASCH solutions have a lower viscosity as compared to the bentonite control. The viscosity of each fluid is attributed to the mode of clay platelet interaction as brought about by surface charges, either through the addition of Fe_2O_3 NP or through the introduction of intercalated clay platelets.

The addition of ICH particles into the bentonite solution promoted attraction of positives charges on the surfaces of the hybrid clay platelet with the negative charges of the bentonite platelets. This attraction was between the edge of the ICH particle and the face of a non-hybrid bentonite clay platelet (E–F) as well as attraction between the face of the hybrid iron-oxide clay platelet and the face of the non-hybrid clay platelet (face-to-face (F–F) flocculation), resulting in the formation of a rigid 3D clay platelet network. This network can be seen in Fig. 4(b) and is graphically illustrated in the panel below image (b). This rigid clay platelet network increased the viscosity of the fluid and the resistance to flow as evidenced through an increase in the yield stress by nearly one order of magnitude, as compared to the bentonite control (Jung et al., 2011).

In contrast, embedded Al_2O_3–SiO_2 nanoparticles promoted negative charges at the edge of the ASCH particle, increasing the negativity of the net charge of the hybrid particle. This resulted in the generation of repulsive forces between hybrid ASCH and bentonite clay platelets in the fluid system. This strong electrostatic repulsion prevented coagulation and the subsequent formation of a strong clay platelet network. This formation is illustrated in Fig. 4(c) and is graphically illustrated in the panel below image (c). Even under HTHP conditions, where typical disassociation of Na^+ ions within the bentonite leads to flocculation and an increase in the viscosity of the fluid system (Laribi et al., 2006, Ramos-Tejada et al., 2001 and Tombáccz and Szekeres, 2004), ASCH prevented this flocculation of clay platelets, yielding a well-dispersed fluid, as evidenced by a reduction in the flow stress as compared to the bentonite control (Jung et al., 2011).

The addition of Fe_2O_3 NP induced heterocoagulation, or the assembly of different types of clay particles in a collective behavior through the introduction of an electrolyte to the clay solution, promoting a rigid structure (Brandenburg and Lagaly, 1988, Lagaly, 1989 and Tombácz et al., 2001) and a stronger gel formation (Callaghan and Ottewill, 1974 and Jung et al., 2011). This yielded an increase in viscosity and flow stress under LTLP conditions in comparison to the bentonite control, indicating flocculation through edge-to-edge (E–E) and edge-to-face (E–F) associations. Flocculation was induced due to the imbalance of electrical charges on the surface of the clay platelet, as seen in the SEM images in Fig. 4(d) and (e) and graphically illustrated in the panel between images (d) and (e). The Fe_2O_3 NP exhibit positive charges in an aqueous environment and will attract the negatively charged faces of the bentonite clay platelets. Under HTHP conditions, the viscosity and flow stress of the Fe_2O_3 solutions minimally increased in comparison to the bentonite control, indicating flocculation via the same associations between edge and face as seen under LTLP conditions (Jung et al., 2011).

It is evident the addition of 0.5 wt% 3 nm Fe_2O_3 NP had a more substantial effect on viscosity and flow stress of the bentonite solution than did the 30 nm counterpart. The viscosity increased 194% and 40% in comparison to the bentonite control for the 3 nm samples under LTLP and HTHP conditions, respectively, in comparison to the 66% and 5% increase in LTLP and HTHP viscosities of the 30 nm samples. The flow stresses increased on the same order of magnitude as the viscosities for both samples under both conditions in comparison to the control (Jung et al., 2011). The reason for an increase in viscosity and flow stress is how the NP associate with the clay platelets, with themselves, and the subsequent interaction between clay platelets, as evidenced by zeta potential (ZP) measurements.

The 3 and 30 nm Fe_2O_3 NP with concentration of 5 mg/1 mL in DI water with a pH of 8.2-8.6 have ZP values of 15.22 ± 0.78 and 17.44 ± 0.42 respectively, indicating unstable colloidal solutions with a high potential of aggregation (in the chemistry definition). Stable colloidal solutions typically have absolute ZP values greater than 30. In 5.0 wt% bentonite solution, the 0.5 wt% 3 and 30 nm Fe_2O_3 solutions have ZP values of −23.50±1.73 and −7.53±2.00 mV and pH values of 8.64 and 8.82. In comparison, the ZP of the 5.0 wt% bentonite solution was found to be −7.06±2.64. Typically, there is a large negative charge generated in aqueous Mt solutions due to substitution of Si^{4+} for Al^{3+}, Mg^{2+}, and Fe^{2+} in the Mt aluminosilicate layers. However, based on the magnitude of the ZP for the 5.0 wt% bentonite solution, there is not enough electrostatic repulsion within the colloidal solution to keep it well-dispersed, hence the onset of flocculation. The larger magnitude of the ZP of the 3 nm Fe_2O_3 NP solution indicates greater stability than the bentonite control and 30 nm Fe_2O_3 solution.

The relative stability of the 3 nm Fe_2O_3 NP solution is due to the larger surface area to volume ratio of the 3 nm Fe_2O_3 NP as compared to the 30 nm Fe_2O_3 NP. A larger surface area to volume ratio leads to stronger electrical attraction with the negatively charged face of the bentonite clay platelet. Furthermore, for the same weight percent, there are more 3 nm Fe_2O_3 NP participating in attraction with the betonite clay platelet. Additionally, the 3 nm Fe_2O_3 NP experience aggregation between NP due to electrostatic attraction. The mutual attraction between NP and clay platelet surfaces results in a heterocoagulated structure as seen in bottom middle panel of Fig. 4. The 3 nm Fe_2O_3 heterocoagulated bentonite clay platelets are positively charged, repelling like structures and platelets within the solution and thus the solution exhibits better dispersion (higher ZP value) and stability than the 30 nm Fe_2O_3 and bentonite solutions.

The pH values of the 3 and 30 nm Fe_2O_3 NP solutions are 8.64 and 8.82, respectively. Lower pH values (<7.5) lead to E–E and E–F flocculation due to the minimization of surface charge imbalance (Stawinksi et al., 1990). Increasing the pH leads to an increase in F–F associations, where the clay platelets aggregate, forming a linked structure (Vali and Bachmann, 1988). This linked structure is able to trap more water between the layers, increasing the viscosity, yield stress and subsequently decreasing fluid filtration volume. The addition of the 3 and 30 nm Fe_2O_3 NP did not substantially affect the pH of the solution as to increase E–E or E–F flocculation, nor did they induce F–F flocculation. Thus, the heterocoagulated structure increased the viscosity and yield stress, and as will be presented in the following section, LTLP fluid filtration loss.

Increasing the temperature and pressure of the drilling fluid changes the viscosity, and thus the rheological properties and subsequent fluid filtration performance. It has been proposed that elevated temperatures induce the dissociation of Na^+ from the surface of bentonite, leading to an increase in viscosity (Laribi et al., 2006, Ramos-Tejada et al., 2001 and Tombáccz and Szekeres, 2004), as evidenced by Fig. 3(a) and (b). For example, Samples A–E experienced an 18.9-, 9.0-, 11.9-, 5.5-, and 5.1-fold increase in apparent viscosity when tested under

HTHP as compared to LTLP conditions, respectively. The effect of pressure, although acting to increase the viscosity via increased particle interaction, is not as great as temperature (Alderman et al., 1988 and Briscoe et al., 1994). From prior work, it can be seen that the effect of pressure is less than temperature on the viscosity of betonite, ICH and ASCH solutions (Jung et al., 2011).

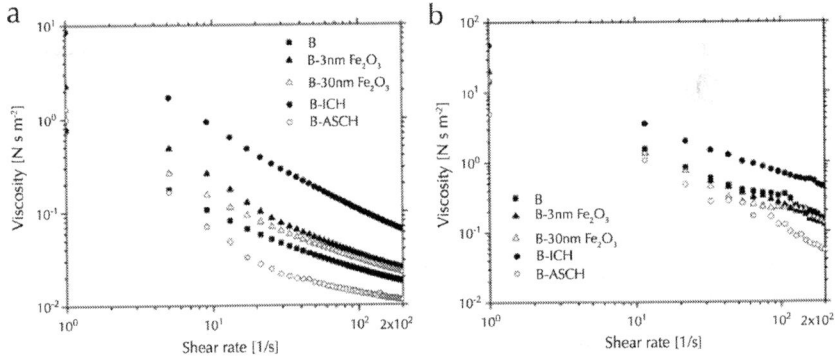

Figure 3: Viscosity versus shear rate of various drilling fluids under (a) LTLP and (b) HTHP conditions.

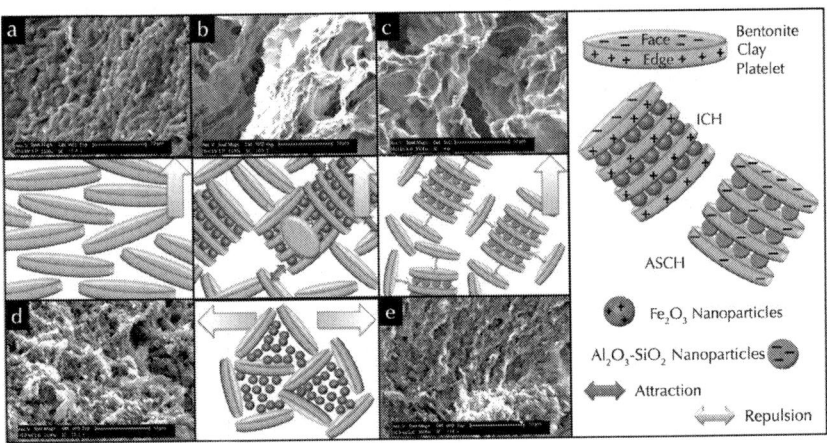

Figure 4: SEM of LTLP filter cakes: (a) Sample A, (b) Sample D, (c) Sample E, (d) Sample B, (e) Sample C, with corresponding clay platelet interaction illustrations.

Fluid Filtration Properties

LTLP

Figure 5 depicts the behavior of the entirety of the bentonite solutions fluid filtrate loss per unit square-root of time in comparison to the control under LTLP conditions. Table 3 summarizes the filtration volumes of the fluids under both LTLP and HTHP conditions. The fluid filtrate loss is plotted against the square root of time for the solution of Darcy's Law (Eq. (2)) is in the form of Eq. (3) and additionally, after the spurt loss, the fluid filtrate loss becomes linear with respect to the square root of filtration time. Fitting the filtration volume versus times as per the power law yields exponent values in the range of 0.49–0.51 as seen in Table 4, which do not reflect the idealized value of 0.5 from Darcy's Law.

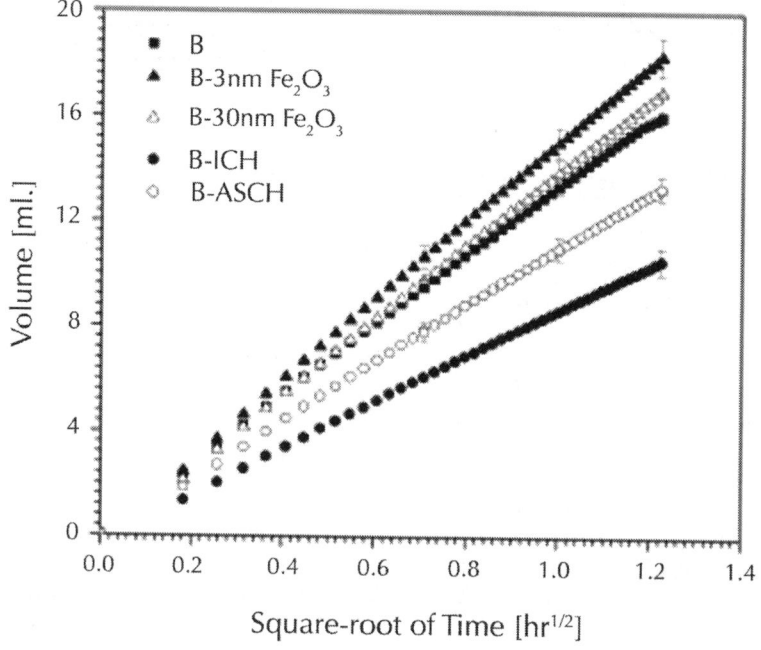

Figure 5: Cumulative LTLP fluid filtration volumes as a function of square-root of time.

Table 3: LTLP and HTHP fluid filtrate volumes V (all volumes are in mL and are reported at time in minutes)

Testing condition		Sample A	Sample B	Sample C	Sample D	Sample E
LTLP	V_{30}	9.6	10.7	9.8	6.1	7.8
	% change	–	+11.5	+2.1	−36.5	−18.8
	V_{60}	13.4	12.6	13.9	8.6	11.0
	% change	–	−6.0	+3.7	−35.8	−17.9
	V_{90}	16.0	18.3	17.0	10.6	13.3
	% change	–	+14.4	+6.3	−33.8	−16.9
HTHP	V_{30}	41.0	29.7	31.4	21.6	32.9
	% change	–	−27.6	−23.4	−47.3	−19.8
	V_{60}	51.6	40.4	42.3	29.6	39.7
	% change	–	−21.7	−18.0	−42.6	−23.1
	V_{90}	55.5	47.5	49.3	35.9	40.9
	% change	–	−14.4	−11.2	−35.3	−26.3

Table 4: Exponential fit of filtration data under LTLP and HTHP conditions

Sample	LTLP	HTHP
A	0.49±1e−4	0.40±0.127
B	0.50±1e−4	0.50±0.078
C	0.51±1e−4	0.32±0.105
D	0.51±1e−4	0.48±0.101
E	0.49±1e−4	0.44±0.034

The ICH and ASCH fluids exhibit the least fluid loss. The ICH and ASCH solutions exhibited a 37% and 19%, 36% and 18%, and 34% and 17% reduction in 30-, 60- and 90-min fluid filtration volumes as compared to the 5.0 wt% bentonite, respectively. Darcy's Law describes fluid filtrate volumes dV as a function of time dt, cross-sectional area A, permeability κ, pressure differential ΔP, viscosity η and filter cake thickness tc, which is expressed as

$$\frac{dV}{dt} = \frac{\kappa \Delta PA}{\eta t_c}.$$

(2)

The solution to Darcy's Law as a function of the aforementioned variables is expressed as

$$V_f = A\sqrt{\frac{2t\kappa\Delta P}{\eta}\left(\frac{f_{sc}}{f_{sm}} - 1\right)},$$

(3)

where f_{sm} is the volume fraction of solids within the mud and f_{sc} is the volume fraction of solids within the cake. To calculate the permeability of each solution's filter cake, the following assumptions were made; the cross-sectional area is constant at 45 cm²; the pressure difference is constant at 6.9 bar; the viscosity of the fluid of interest is water taken at 25 °C; the filter cake thickness, although a function of time, is assumed constant and is taken as the reported final measured thickness as seen in Table 5.

Table 5: Permeability κ calculated from Darcy's Law from filter cake thicknesses tc under LTLP conditions

Sample	tc(cm)	10^{-3}(mD)
A	3.82	3.33
B	5.55	5.53
C	4.53	4.18
D	4.82	2.77
E	3.54	2.57

It is evident from Table 5 that the permeability of the additive 3 and 30 nm Fe_2O_3 filter cakes prepared under LTLP conditions are 66% and 26% higher than the bentonite control, respectively. As observed from Fig. 5, the additive Fe_2O_3 NP bentonite solutions increased fluid loss as compared to the control, which is attributed to the increase in permeability of the filter cake due to the heterocoagulated structure. Additionally, an increase in filtrate volume would yield an increase in filter cake thickness as seen in Table 5. While the size of the Fe_2O_3

NP increased from 3 to 30 nm, the solution exhibited a small decrease in fluid filtrate volume which is attributed to a slight decrease in filter cake permeability. The heterocoagulated formation, which formed a strong clay platelet network, did not experience compaction during filtration testing, thus the high permeability filter cake. As the particle size increased, the degree of heterocoagulation decreased, reducing the filter cake permeability and filtrate volume.

Table 6: Conversion between metric and U.S. customary units

Physical quantity	Metric units	U.S. customary units
Length	1 m	39.3701 in
Pressure	1 bar	14.5038 psi
Temperature	1 °C	$(° F{-}32) \cdot \dfrac{5}{9}$
Viscosity	1 Pa s	24.1909 pound/foot-hour
Volume	1 mL	0.0338 fluid ounce

The ICH and ASCH solutions produced filter cakes of lesser permeability which reflect a 17% and 23% reduction in in comparison to the bentonite control, as provided in Table 5. The embedded Fe_2O_3 NP promoted positive charges at the edge of the particles, bringing about strong attractive forces between the positively charged edge of ICH particles and negatively charged face of clay particles, exhibiting a house-of-cards structure. Strongly cross-linked formations provide more rigid structure which exhibit apparent low permeability, explaining a decrease in the filtrate volumes. In contrast, embedded Al_2O_3–SiO_2 NP promote negative charges at the edge of the particle, leading to a repulsive force between ASCH and bentonite clay platelets. This electrostatic repulsion prevented clay platelet coagulation, and additionally maintained a well-dispersed fluid. The permeability decreased due to subsequent compaction of the well-dispersed, non-coagulated platelets, as evidenced by the filter cake thickness which was less than the bentonite control, yielding a reduction in filtrate volume.

The effect of intercalation and additive NP is illustrated by the SEM images in Fig. 4. Figure 4(a) shows the control bentonite filter cake, whereas Fig. 4(b) and (c) shows the ICH and ASCH filter cakes. The ICH filter cake exhibits predominant F–F association through electrostatic attraction between the face of the hybrid iron-oxide clay platelet and the face of the non-hybrid clay platelet. This resulted in the formation of a rigid 3D clay platelet network, which increased the resistance to flow and decreased the permeability. The ASCH filter cake as seen in Fig. 4(c) is the result of a strong cross-linked clay platelet network formation that is a result of strong electrostatic repulsion between ASCH and bentonite clay platelets. This strong network formation reduced the porosity and subsequent permeability.

Figure 4(d) and (e) shows the SEM images, concurrent with literature (van Olphen, 1964), and the schematic illustration representing hetercoagulated formations of Fe_2O_3 NP and clay platelets. Randomly shaped and empty spaces are created between the particles and thus filtration easily occurs through these pores, leading to higher permeability values and fluid filtrate volumes.

HTHP

Figure 6(a) and (b) depicts the behavior of the hybrid (ICH and ASCH) and 3 and 30 nm additive Fe_2O_3 bentonite solutions fluid filtrate loss per unit square-root of time under HTHP conditions, respectively, along with the results being summarized in Table 3.

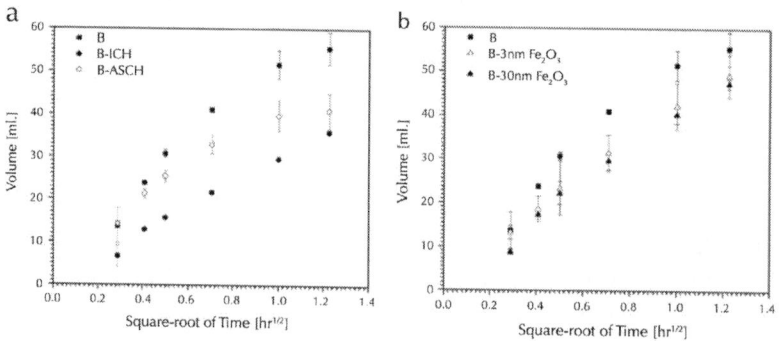

Figure 6: Cumulative (a) hybrid and (b) additive HTHP fluid filtration volumes as a function of square-root of time.

The ICH and ASCH added fluids exhibit the least fluid loss in comparison to the bentonite control, with a reduction in fluid filtrate volumes by 47% and 20% for the 30-min, 43% and 23% for the 60-min and 35% and 26% for the 90-min tests, respectively. With an increase in temperature, the ICH filter cake retained the strong positive charges, which attracted the negative charges on the faces of the bentonite platelets, thus creating a rigid house-of-cards structure with a high viscosity. The house-of-cards structure allowed for the formation of a filter cake with low-permeability characteristics. The permeability of the ICH filter cake was decreased due to reduced pore sizes, which is attributed to the collapse of the hybrid and non-hybrid clay platelets via the strong attraction between hybrid and bentonite platelets faces (F–F). This low porosity and low permeability filter cake exhibited low filtrate volume in comparison to the bentonite control.

The negative charge of the ASCH clay platelets repulsed the bentonite clay platelets and acted as a deflocculant, promoting clay platelet dispersion within the solution at elevated temperatures (Bourgoyne et al., 1986 and Jung et al., 2011). This dispersion resulted in the production of a low porosity and low permeability filter cake, resulting in lesser fluid filtrate volume as compared to the bentonite control.

Figure 6(b) depicts the filtration volumes of bentonite fluids containing Fe_2O_3 additive NP under HTHP conditions. Contrasting to the indications in Fig. 4, under HTHP conditions, Fe_2O_3 additive NP solutions showed a decrease in fluid filtration volume as compared to the control bentonite fluid. Elevated temperatures induce flocculation of clay platelets and the dissociation of Na^+ from the surface of bentonite; as Na^+ concentrations increase, pre-existing network formations collapse and the solution became flocculated, creating a high porosity and high permeability filter cake, as evidenced by a 330%, 286% and 246% increase in the 30-, 60- and 90-min HTHP filtration volumes of the 5.0 wt% bentonite solution as compared to LTLP. However, the Fe_2O_3 NP replaced the disassociated Na^+ and allowed the clay platelet to retain a positive charge along the edge; this retention of positive edge charge kept the clay platelets dispersed and deflocculated. Thus, the deflocculated solution exhibits a low porosity and low permeability filter cake, thus less filtrate volume in comparison to the bentonite control.

The filter cake permeability and subsequent fluid filtration are influenced by the clay platelet interaction under HTHP conditions; ICH added solutions exhibit rigid networks and low permeability filter cakes due to the strong network formation between positively charged hybrid platelets and negatively charged bentonite edges, thus low filtration volumes; ASCH added solutions exhibit good dispersion due to repulsion generated between negatively charged hybrid particles and negatively charged bentonite faces, thus minimal clay platelet network formation and low permeability filter cakes; additive Fe_2O_3 solutions are deflocculated due to the replacement of dissociate Na^+ cations with Fe_2O_3 NP, leading to a solution that yields low permeability filter cakes.

This work illustrates how modifying the surface charge of clay-based drilling fluids via intercalation and NP addition can alter the rheological properties and fluid filtration characteristics of conventional drilling muds under LTLP and HTHP conditions.

CONCLUSIONS

The rheological properties and fluid filtration performance of low solid content bentonite fluids containing iron-oxide (Fe_2O_3) additives and nanoparticle (NP) intercalated clay hybrids under low-temperature low-pressure (LTLP: 25 ° C, 6.9 bar) and high-temperature high-pressure (HTHP: 200 ° C, 70 bar) conditions were investigated.

The iron-oxide clay hybrid (ICH) added hybrid solution exhibited less fluid filtrate as compared to the bentonite control under LTLP and HTHP conditions. ICH added solutions exhibited strong cross-linked and coagulated platelet networks under LTLP and HTHP conditions. The cross-linked platelet network was less sensitive to pressure and temperature which resulted in less permeable filter cakes, reducing fluid filtrate volumes under LTLP and HTHP conditions.

The LTLP and HTHP fluid filtrate volumes of the aluminosilicate clay hybrid (ASCH) added hybrid solutions were less than the bentonite control due to the reduced permeability of the filter cakes. The effect on permeability is attributed to the strong electrostatic repulsion between the hybrid particles and clay platelets which provided good dispersion and prevented coagulation and flocculation.

The pure addition of 0.5 wt% 3 and 30 nm Fe_2O_3 NP in 5.0 wt% bentonite solutions increased the fluid filtrate volume under LTLP conditions as compared to the 5.0 wt% bentonite control. Fe_2O_3 NP promoted heterocoagulation of clay platelets which resulted in the formation of a permeable filter cake during filtration testing. Furthermore, the pure addition of Fe_2O_3 NP to the bentonite solution tested under HTHP reduced fluid filtrate volumes as compared to the bentonite control. The fluid filtrate volume decreased at elevated temperatures because Fe_2O_3 NP replaced dissociated Na^+ cations, deflocculating the solution which yielded a low permeability filter cake.

SEM images of the Fe_2O_3 additive solutions showed rigid, porous structures under LTLP conditions and less porous structures under HTHP conditions. SEM images of the ICH added solution filter cakes showed a semi-porous cross-linked structure under both LTLP and HTHP conditions, and those of the ASCH added solutions filter cakes showed random, coagulated structures under both LTLP and HTHP conditions.

ACKNOWLEDGEMENTS

This work was supported by the Department of Energy.

REFERENCES

1. Alderman, N.J., Gavignet, A., Cuillot, D., Maitland, G.C., et al., 1988. Hightemperature high-pressure rheology of water-based muds. In: SPE Annual Technical Conference and Exhibition, Society of Petroleum Engineers.

2. Arthur, K., Peden, J., 1988. The evaluation of drilling fluid filter cake properties and their influence on fluid loss. In: International Meeting on Petroleum Engineering, Society of Petroleum Engineers, Tianjin, China.

3. Baird, J.C., Walz, J.Y., 2007. The effects of added nanoparticles on aqueous kaolinite suspensions. II. Rheological effects. J. Colloid Interface Sci. 306 (2), 411–420.

4. Belani, A., Orr, S., 2008. Management—a systematic approach to hostile environments. J. Pet. Technol. 60 (7).

5. Bourgoyne, Adam T., Keith K. Millheim, Martin E. Chenevert, and F. S. Young. "Applied drilling engineering." (1986). This is from SPE Textbook Series, Vol. 2.

6. Brandenburg, U., Lagaly, G., 1988. Rheological properties of sodium montmorillonite dispersions. Appl. Clay Sci. 3 (3), 263–279.

7. Briscoe, B.J., Luckham, P.F., Ren, S.R., 1994. The properties of drilling muds at high pressures and high temperature. Philos. Trans. R. Soc. Lond. Ser. A: Phys. Eng. Sci. 348 (1687), 179–207.

8. Callaghan, I.C., Ottewill, R.H., 1974. Interparticle forces in montmorillonite gels. Faraday Discuss. Chem. Soc. 57, 110–118.

9. Chenevert, M. Huycke, J., 1991. Filter Cake Structure Analysis Using the Scanning Electron Microscope. Technical Report. Society of Petroleum Engineers.

10. Darley, Henry CH, and George Robert Gray. Composition and properties of drilling and completion fluids. Gulf Professional Publishing, 1988.

11. Hartmann, L., Özerler, M., Marx, C., Neumann, H.-J., 1988. Analysis of mudcake structures formed under simulated borehole conditions. SPE Drill. Eng. 3 (4), 395–402.

12. Huang, T., Crews, J.B., 2008. Nanotechnology applications in viscoelastic surfactant stimulation fluids. SPE Prod. Oper. 23 (4), 512–517.

13. Huang, T., Crews, J.B., Agrawal, G., 2010. Nanoparticle psuedocrosslinked micellar fluids: optimal solution for fluid-loss control with internal breaking. In: SPE International Symposium and Exhibition on Formation Damage Control, Society of Petroleum Engineers.

14. Jung, Y., Son, Y.-H., Lee, J.-K., Phuoc, T.X., Soong, Y., Chyu, M.K., 2011. Rheological behavior of clay nanoparticle hybrid-added bentonite suspensions: specific role of hybrid additives on the gelation of clay-based fluids. ACS Appl. Mater. Interfaces 3 (9), 3515–3522.

15. Kelessidis, V.C., Tsamantaki, C., Pasadakis, N., Repouskou, E., Hamilaki, E., 2006. Permeability, porosity and surface

CHAPTER 7

F. Ouachtari, A. Rmili, B. Elidrissi, A. Bouaoud, H. Erguig and P. Elies, "Influence of Bath Temperature, Deposition Time and S/Cd Ratio on the Structure, Surface Morphology, Chemical Composition and Optical Properties of CdS Thin Films Elaborated by Chemical Bath Deposition," Journal of Modern Physics, Vol. 2 No. 9, 2011, pp. 1073-1082. doi: 10.4236/jmp.2011.29131.

CHAPTER 8

Doelle, K. , Le, A. , Amidon, T. and Bujanovic, B. (2014) Use of Poly-Lactic Acid (PLA) to Enhance Properties of Paper Based on Recycled Pulp. Advances in Chemical Engineering and Science, 4, 347-360. doi: 10.4236/aces.2014.43038.

CHAPTER 9

Rosales, I. and Martínez, H. (2014) Defects Interaction on the Mechanical Properties during Transition Formation of (Mo, Cr) 3Si Intermetallic Alloys. Journal of Materials Science and Chemical Engineering, 2, 64-70. doi:10.4236/msce.2014.211009.

CHAPTER 10

Matthew M. Barry, Youngsoo Jung, Jung-Kun Lee, Tran X. Phuoc, Minking K. Chyu, Fluid filtration and rheological properties of nanoparticle additive and intercalated clay hybrid bentonite drilling fluids, Journal of Petroleum Science and Engineering, Volume 127, March 2015, Pages 338-346, ISSN 0920-4105, http://dx.doi.org/10.1016/j.petrol.2015.01.012.

Index